T0178541

Lecture Notes in Computer Science 13262

More information about this series at https://link.springer.com/bookseries/558

Cameron Browne · Akihiro Kishimoto ·
Jonathan Schaeffer (Eds.)

Advances in Computer Games

17th International Conference, ACG 2021
Virtual Event, November 23–25, 2021
Revised Selected Papers

Springer

Editors
Cameron Browne 🆔
Maastricht University
Maastricht, The Netherlands

Akihiro Kishimoto 🆔
IBM Research - Tokyo
Tokyo, Japan

Jonathan Schaeffer
University of Alberta
Edmonton, AB, Canada

ISSN 0302-9743   ISSN 1611-3349 (electronic)
Lecture Notes in Computer Science
ISBN 978-3-031-11487-8   ISBN 978-3-031-11488-5 (eBook)
https://doi.org/10.1007/978-3-031-11488-5

This Springer imprint is published by the registered company Springer Nature Switzerland AG
The registered company address is: Gewerbestrasse 11, 6330 Cham, Switzerland

Preface

This book collects the papers presented at the 17th Advances in Computer Games conference (ACG 2021) which took place during November 23–25, 2021. The conference was held online in 2021 for the first time in its 47 year history due to the COVID-19 pandemic and subsequent travel restrictions. This was a truly international effort as the conference was primarily hosted from Maastricht University, the Netherlands, through a Zoom stream provided by the University of Alberta, Canada, with assistance from IBM Research - Tokyo, Japan.

The Advances in Computer Games conference series is a major international forum for researchers and developers interested in all aspects of artificial intelligence and computer game playing. Earlier conferences took place in London (1975), Edinburgh (1978), London (1981, 1984), Noordwijkerhout (1987), London (1990), Maastricht (1993, 1996), Paderborn (1999), Graz (2003), Taipei (2005), Pamplona (2009), Tilburg (2011), Leiden (2015, 2017) and Macao (2019). For the past 20 years, the conference has been held every second year, alternating with the Computer and Games conference.

A total of 34 papers were submitted to this conference. One was later withdrawn and the remaining 33 papers were each reviewed by three reviewers. A total of 22 papers were accepted for presentation.

The online nature of this year's conference offered some benefits in that conference registration could be made free for the first time, resulting in a record participation for this event with 399 registered participants. It also provided logistical challenges in devising a schedule that worked for as many attendees as possible over a wide range of time zones.

The themes for this year's ACG conference were specifically widened to include video game research in addition to the usual traditional/mathematical games research. The goal was to broaden the conference focus to encourage new researchers to participate in International Computer Games Association (ICGA) events. The four papers presented in Session 5: Player Modelling represent the results of this initiative.

The ACG 2021 program consisted of three keynote speeches and six regular paper sessions. The keynote talks were from world-class researchers David Silver and Michael Bowling, along with veteran computer chess program creators Larry Kaufman and Mark Lefler. All papers and presentation videos can be accessed at the ACG 2021 web site.[1]

Session 1: Learning in Games

The opening session, chaired by Todd Neller, collected three papers focussed on machine learning – especially deep learning – for specific games. These included C. Yi and T. Kaneko on "Improving Counterfactual Regret Minimization Agents Training in the Card Game Cheat", B. Doux, B. Negrevergne, and T. Cazenave on "Deep Reinforcement

---

[1] https://icga.org/?page_id=3328.

Learning for Morpion Solitaire", and L. G. Heredia and T. Cazenave on "Expert Iteration for Risk".

Session 2: Search in Games

This session, chaired by Michael Hartisch, presented new search methods and enhancements of existing search methods for a range of games. The papers presented were N. Fabiano and T. Cazenave on "Sequential Halving Using Scores", T. Cazenave, J. Sentuc, and M. Videau on "Cosine Annealing, Mixnet and Swish Activation for Computer Go", G. Moskowitz and V. Ponomarenko on "A Heuristic Approach to the Game of Sylver Coinage", and A. Pálsson and Y. Björnsson on "Evaluating Interpretability Methods for DNNs in Game-Playing Agents".

Keynote: Artificial Intelligence Goes All-In: Computers Playing Poker

The first keynote speaker, Michael Bowling from the University of Alberta's Computer Poker Research Group and Google DeepMind, was introduced by Jonathan Schaeffer. This talk described the development of the world's first superhuman Poker bots.

Session 3: Solving Games

Kazuki Yoshizoe chaired this session on solving, or at least providing more complete complexity analyses, of some simple games. This included S. Tanaka, F. Bonnet, S. Tixeuil, and Y. Tamura on "Quixo is Solved", J. Uiterwijk on "Solving Bicoloring-Graph Games on Rectangular Boards – Part 1: Partisan Col and Snort" and "Part 2: Impartial Col and Snort", and R. Hayward, R.A. Hearn, and M. Jamshidian on "BoxOff is NP-Complete".

Keynote: 54 Years of Progress in Computer Chess

Larry Kaufman and Mark Lefler, introduced by Jaap van den Herik, presented a personal account of key developments in computer chess over the last half century, through their own experiences in the field.

Session 4: Chess Patterns

This session on chess patterns, chaired by Tristan Cazenave, explored effective representations of chess for AI search. The papers included M. Bizjak and M. Guid on "Automatic Recognition of Similar Chess Motifs", R. Haque, T.H. Wei, and M. Müller on "On the Road to Perfection? Evaluating Leela Chess Zero Against Endgame Tablebases", and D. Gomboc and C. Shelton on "Chess Endgame Compression via Logic Minimization".

Session 5: Player Modelling

Matthew Stephenson chaired Session 5 on Player Modelling which collected the video games papers accepted for the conference. These included K. Fujihira, C.-H. Hsueh, and K. Ikeda on "Procedural Maze Generation with Difficulty from Human Players' Perspectives", H.-J. Chang, C. Yueh, G.-Y. Fan, T.-Y. Lin, and T.-S. Hsu on "Opponent Model Selection Using Deep Learning", G. Guglielmo, I.F. Peradajordi and M. Klincewicz on "Deep Learning to Detect Facial Markers of Complex Decision Making", and A. Gunes, F. Kavum and S. Sariel on "Player Modeling Using Event-Trait Mapping Supported by PCA".

Keynote: AlphaZero Fundamentals

The third keynote speaker was David Silver, leader of Google DeepMind's machine learning group and lead researcher on the successful AlphaGo, AlphaZero, and AlphaStar programs. David, introduced by Martin Müller, described the inner workings of AlphaZero, and how existing search methods were adapted to produce the spectacular results obtained.

Session 6: Game Systems

This session, chaired by Spyridon Samothrakis, featured four papers on game systems, especially the Ludii general game system. The papers included M. Stephenson, E. Piette, D.J.N.J. Soemers, and C. Browne on "Automatic Generation of Board Game Manuals", "Optimised Playout Implementations for the Ludii GGS", and "General Board Geometry" (in various orders of authorship) in addition to M. Goadrich and C. Shaddox on "Quantifying the Space of Hearts Variants".

Acknowledgements

The organization of ACG 2021 was partly supported by the European Research Council as part of the Digital Ludeme Project (ERC CoG #771292). The online technology was provided by the University of Alberta's AI4Society. Thank you to Nicolás Arnáez and Eleni Stroulia for their tremendous support.

November 2021   Cameron Browne
                                                               Akihiro Kishimoto
                                                               Jonathan Schaeffer

Organization

Organizing Committee

Cameron Browne	Maastricht University, The Netherlands
Akihiro Kishimoto	IBM Research - Tokyo, Japan
Jonathan Schaeffer	University of Alberta, Canada

ICGA Executive

Tristan Cazenave	LAMSADE, Université Paris Dauphine PSL, CNRS, France
Hiroyuki Iida	Japan Advanced Institute of Science and Technology, Japan
David Levy	Independent, UK
Jonathan Schaeffer	University of Alberta, Canada
Jaap van den Herik	University of Leiden, The Netherlands
Mark Winands	Maastricht University, The Netherlands
I-Chen Wu	National Yang Ming Chiao Tung University, Taiwan

Program Committee

Yngvi Björnsson	Reykjavik University, Iceland
Bruno Bouzy	Paris Cité University, France
Cameron Browne	Maastricht University, The Netherlands
Tristan Cazenave	LAMSADE, Université Paris Dauphine – PSL, CNRS, France
Lung-Pin Chen	Tunghai University, Taiwan
Siang Yew Chong	University of Nottingham, Malaysia
Chao Gao	University of Alberta, Canada
Reijer Grimbergen	Tokyo University of Technology, Japan
Michael Hartisch	University of Siegen, Germany
Ryan Hayward	University of Alberta, Canada
Chu-Hsuan Hsueh	Japan Advanced Institute of Science and Technology, Japan
Hiroyuki Iida	Japan Advanced Institute of Science and Technology, Japan
Eric Jacopin	CREC Saint-Cyr, France
Nicolas Jouandeau	Paris 8 University, France

Tomoyuki Kaneko	University of Tokyo, Japan
Akihiro Kishimoto	IBM Research - Tokyo, Japan
Jakub Kowalski	University of Wroclaw, Poland
Sylvain Lagrue	Université de Technologie de Compiègne, France
Diego Perez Liebana	Queen Mary University of London, UK
Shun-Shii Lin	National Taiwan Normal University, Taiwan
Richard Lorentz	California State University, USA
Martin Müller	University of Alberta, Canada
Todd Neller	Gettysburg College, USA
Mark Nelson	American University, USA
Éric Piette	Maastricht University, The Netherlands
Mike Preuss	Universiteit Leiden, The Netherlands
Abdallah Saffidine	University of New South Wales, Australia
Spyridon Samothrakis	University of Essex, UK
Jonathan Schaeffer	University of Alberta, Canada
Dennis Soemers	Maastricht University, The Netherlands
Matthew Stephenson	Maastricht University, The Netherlands
Nathan Sturtevant	University of Alberta, Canada
Ruck Thawonmas	Ritsumeikan University, Germany
Michael Thielscher	University of New South Wales, Australia
Jonathan Vis	Universiteit Leiden, The Netherlands
Ting Han Wei	University of Alberta, Canada
Mark Winands	Maastricht University, The Netherlands
I-Chen Wu	National Yang Ming Chiao Tung University, Taiwan
Shi-Jim Yen	National Dong Hwa University, Taiwan
Kazuki Yoshizoe	Kyushu University, Japan

Contents

Learning in Games

Improving Counterfactual Regret Minimization Agents Training in Card Game Cheat Using Ordered Abstraction

Cheng Yi[✉] and Tomoyuki Kaneko

Graduate School of Arts and Sciences, The University of Tokyo, Tokyo, Japan
yi-cheng199@g.ecc.u-tokyo.ac.jp, kaneko@acm.org

Abstract. Counterfactual Regret Minimization (CFR) has been one of the most famous iterative algorithms to learn decent strategies of imperfect information games. Vanilla CFR requires traversing the whole game tree on every iteration, which is infeasible for many games, especially as a number of them require infinite steps due to repeated game states. In this paper, we introduce an abstraction technique called Ordered Abstraction to enable us to train using a much smaller and simpler version of the game by limiting the depth of the game tree. Our experiments were conducted in an imperfect information card game called Cheat, and we introduce the notion of "Health Points" a player has in each game to make the game length finite and thus easier to handle. We compare four variants of CFR agents, evaluate how the results from smaller games can improve the training in larger ones, and show how Ordered Abstraction can help us increase the learning efficiency of specific agents.

Keywords: Imperfect Information Games · Counterfactual Regret Minimization · Abstraction Technique · Curriculum Learning

1 Introduction

In artificial intelligence research, we often see games as our challenging problems and solving them represents research breakthroughs. There are two kinds of games: perfect information and imperfect information. In imperfect information games, such as Bridge, Mahjong, and most poker games, players do not know the complete game state. The hidden information of the play is what makes imperfect information games more challenging. The Nash equilibrium (NE), which is an important concept in game theory, is a strategy profile where no player can achieve a better result through converting their strategy unilaterally. Therefore, the goal of most research is to reach or approximate the NE of games.

Counterfactual Regret Minimization (CFR) has become one of the most famous and widely-used algorithms when dealing with imperfect information games. In 2018, Noam Brown et al. developed an agent called Libratus [3] based on CFR for Limited Texas Hold'em, a widely known and played poker game,

© Springer Nature Switzerland AG 2022
C. Browne et al. (Eds.): ACG 2021, LNCS 13262, pp. 3–13, 2022.
https://doi.org/10.1007/978-3-031-11488-5_1

and further improved it to create Pluribus [4] for a multiplayer no-limit version of the game. Pluribus is the first AI agent to beat top human players in such a complex multiplayer game. The main idea of CFR is to converge to an NE on the basis of the counterfactual regret (how much we are regretful for choosing this action) calculation of every state on the game tree for the players. One of the limitations of this algorithm is that Vanilla CFR requires a traversal of the whole game tree on each iteration, which becomes infeasible when dealing with extremely large games. As a result, researchers have been looking for better ways to deal with infeasibly large games to save computing costs.

In this paper, we introduce a new approach called Ordered Abstraction to create and adjust the training environment of CFR agents to serve our purpose. We limit the total length of the game for simplification and aim at using the results from simpler games in larger games to speed up the iterations and achieve a better result. In the next section we will introduce some background knowledge, and Sect. 3 will cover a number of related previous works. Sections 4 and 5 describe our proposed methods, details of conduction, and the results. In the last section, we summarize the paper and outline the future direction of our research.

2   Background

2.1   Notations and Terminology

Extensive Games and Nash Equilibrium. We followed a standard notation in game theory [5]. A finite extensive form game with imperfect information is composed of the following elements:

- A finite-size set of *players*, $\mathcal{P}$. For player i, $-i$ represents all the players other than i. There is also a Chance player c representing the actions that are not controlled by any player.
- A *history* $h \in \mathcal{H}$ is a node on the game tree, composed of all the information at that exact game state. A *terminal history* $z \in \mathcal{Z} \subseteq \mathcal{H}$ is where there are no more available actions and each player will receive a payoff value for what they have done following the game tree respectively.
- We use A to denote the *action space* of the whole game and $A(h)$ as the set of all *legal actions* for players at history h. If history h' is reached after a player chooses action $a \in A(h)$ at history h, we can write $h \cdot a = h'$.
- An *information set* (infoset) is a set of histories that for a particular player, they cannot distinguish which history they are in between one another. $\mathcal{I}_i$ represents the finite set of all infosets for player i. Inherently, $\forall h, h' \in I, A(h) = A(h') = A(I)$.
- For each player $i \in \mathcal{P}$, there is a payoff function $u_i : \mathcal{Z} \to \mathcal{R}$, and especially in two-player zero-sum games, $u_1 = -u_2$.

In a game, a *strategy* for player i is σ_i, which assigns a distribution over their action space to each infoset of player i, particularly, $\sigma_i^t(I, a)$ for player i maps the infoset I and action $a \in A(I)$ to the probability that player i will specifically choose action a in the infoset I on iteration t. Σ_i denotes the set of all strategies of player i. A strategy profile $\sigma = (\sigma_1, \ldots, \sigma_n)$ is a tuple of all the players' strategies with one entry for each player where σ_{-i} represents the strategies in σ except σ_i. Let $\pi^\sigma(h)$ denote the *reach probability* of reaching the game history h while all the players follow the strategy profile σ. The contributions of player i and all other players to this probability are denoted by $\pi_i^\sigma(h)$ and $\pi_{-i}^\sigma(h)$, respectively.

Informally, the Nash Equilibrium (NE) is the "best" strategy profile, in a sense that a player who follows it can be seen as "no-loss." Here, we will give the formal definition of NE. Let (Σ, u) be a game with n players, where $\Sigma = \Sigma_1 \times \Sigma_2 \times \cdots \times \Sigma_n$ is the set of strategy profiles and $u(\sigma) = (u_1(\sigma), \ldots, u_n(\sigma))$ is its payoff function defined over $\sigma \in \Sigma$. Therefore, the NE can be now expressed as a strategy profile σ^*, in which every player is playing the best response. Formally, a strategy profile $\sigma^* \in \Sigma$ is an *NE* if $\forall i, \sigma_i \in \Sigma_i : u_i(\sigma_i^*, \sigma_{-i}^*) \geq u_i(\sigma_i, \sigma_{-i}^*)$.

2.2   The Game *Cheat*

Cheat is a card game of lying and bluffing while also detecting opponents' deception. One turn of the game consists of two phases *Discard* and *Challenge*. At the beginning of the game, all the cards are well shuffled and dealt to the players as equally as possible. The first discard player is chosen randomly and the "current rank" that all the players share is set to be Ace.

In the *Discard* phase, the discard player discards one or more cards, puts it (them) face down on the table, and makes a claim that includes the number of cards discarded and current rank. Players are supposed to discard cards only of the current rank but they can lie about their cards—either bluffing when they do not hold any correct cards or choosing other cards even if they have the correct ones. Then, in the *Challenge* phase, if any other player thinks the discard player is lying, they can challenge them by saying "Cheat!" When there is a challenge, the last discarded card(s) will be revealed to all players to see whether they are consistent with the claim. If the accused player did lie, they must take the pile of cards on the table back to their hands, otherwise, the challenger takes the pile. If no one challenges, the card(s) remain(s) in the pile. After the *Challenge* phase, we move to the *Discard* phase in the next turn. The current rank increases by one (K is followed by Ace) and the player sitting to the right of the former discard player then discards one or more cards. The first to discard all their cards and survive the last challenge wins the game.

The rule states that cards are discarded and taken back during the game that might lead to repetitions of game states and thus infinite game lengths, which is one of the difficulties we have to overcome when we apply Counterfactual Regret Minimization (CFR). Although our paper focuses on the application of CFR, the incorporation of other techniques would be of interest in the future.

2.3   Counterfactual Regret Minimization

CFR was first proposed in 2008 by Zinkevich et al. [7] where the idea that claims minimizing overall regret can be used to approximate an NE in extensive games with incomplete information was demonstrated and proved. The basic steps of one iteration of Vanilla CFR are as follows. First, CFR keeps a record of the regret values, $R^t(I, a)$, for all actions $a \in A(I)$ (all zeros at the beginning) in each infoset $I \in \mathcal{I}_i$ where t denotes iteration. Second, the values are used to generate strategies, s.t., $\sigma^{t+1}(I, a) \propto \max(R^t(I, a), 0)$. Third, the regret values are updated on the basis of the new strategies. After all iterations, the average strategy $\bar{\sigma}(I, a) = \sum_t \pi^{\sigma^t}_{-i} \sigma^t(I, a)$ obtained by normalizing the overall actions belonging to the action space of this infoset, weighted by counterfactual reach probability, is proved to converge to the best strategy as time tends to infinity.

Vanilla CFR requires traversals of the whole game tree in every iteration. The average game tree size of the original Cheat is massive and possibly infinite, so traversing the entire game tree even once is impossible and the computation is beyond the calculation power of ordinary computers. Another variant called Chance-sampled CFR (CS-CFR) is more common in practice, especially when dealing with poker or card games. We see the results of dealing cards as Chance player's actions, and on each iteration, we only sample the action of the Chance player. In our paper, we focus on CS-CFR and test it among three variants of it in experiments (see details in Sect. 5.1).

3   Related Works

There are many related studies on Cheat. P. Sinclair[1] applied Perfect Information Monte Carlo Tree Search and evaluated the performance against several stochastic rule-based agents.

In another study [6], Neller and Hnath presented an abstraction as well as enhancements on graph search in a dice game called Dudo. They dramatically reduce the time cost of CFR training and were the first one to approximate Nash equilibrium in a full 2-player Dudo. The abstraction they have used and our Ordered Abstraction share a similar structure with natural numbers.

Moreover, there have been many enhancements for CFR to tackle large games. The Blueprint strategy is one of them. It was introduced in Libratus [3] and then improved upon in Pluribus [4]. First, an abstraction of the whole game is defined and the solution to this abstraction is called Blueprint strategy. This strategy only has specific details for the early stage of the game and an approximation for later parts. The approximation will then be refined during the runtime of the game and after the agent learns more about the opponents' actions.

Most of the technique enables us to save time and space for the whole game but remains unchanged at the early stage of the game. In 2015, Brown et al.

---

[1] https://project-archive.inf.ed.ac.uk/ug4/20181231/ug4_proj.pdf.

propounded an algorithm called *simultaneous abstraction and equilibrium finding (SAEF)* [1], which does not rely on any domain knowledge but is only applicable in specific conditions. In 2016, a refined version of SAEF called *Strategy-Based Warm Starting* was introduced [2]. The new method expands the power of SAEF and is capable of skipping the early expensive iterations of the game. Although warm starting and our curriculum learning have a number of similarities, our method is simpler because both the initial strategy and regret are directly transferred while warm starting involves a sophisticated procedure to recover a substitute regret from a given strategy.

4 Ordered Abstraction and Curriculum Learning

To handle a subset of infinite games with CFR, we present Ordered Abstraction. The basic idea is to make a finite variant of an original game by introducing a condition to terminate the game in finite steps. Then, we run CFR to obtain a strategy in this finite variant (abstraction). We hope that the learned strategy would also work well in the original game, but it crucially depends on the design of the abstraction. To remedy such difficulties, we present an effective heuristic of a curriculum learning with an abstraction with numbering as follows:

1. Design a finite variant, G_n, of a game, associated with integer n such that a variant with a smaller n is easier and thus has a stronger restriction (i.e., having a shorter game length and a smaller subset of infosets), and it asymptotically recovers the original game as $n \to \infty$. We assume that for all $n < n'$, $\mathcal{H}^{G_n} \subseteq \mathcal{H}^{G_{n'}}$ and $|\mathcal{I}_i^{G_n}| \leq |\mathcal{I}_i^{G_{n'}}|$ for each player $i \in \mathcal{P}$ and that any non-terminal history is also non-terminal in a larger game, $((\mathcal{H}^{G_n} \setminus \mathcal{Z}^{G_n}) \cap \mathcal{Z}^{G_{n'}}) = \emptyset$. Typically, there are a number of histories that are terminal in G_n and non-terminal in $G_{n'}$ to make variant G_n strictly smaller. We use superscript X^{G_n} to denote property X in variant G_n.
2. Run CFR T iterations in the easiest variant, G_1, to obtain a decent strategy profile $\bar{\sigma}^{t=T,G_1}(I, \cdot)$ and regrets $R^{t=T,G_1}(I, \cdot)$ for each infoset I,
3. Run CFR with variant G_n after completing CFR with variant G_{n-1}, initializing the strategy as well as regret for each infoset by using the results obtained for variant G_{n-1} to speed up learning, i.e., $\sigma^{t=1,G_n}(I, a) \leftarrow \bar{\sigma}^{t=T,G_{n-1}}(I', a)$ and $R^{t=1,G_n}(I, a) \leftarrow R^{t=T,G_{n-1}}(I', a)$ where $I = I'$ for $I \in \mathcal{I}_i^{G_n}$, $I' \in \mathcal{I}_i^{G_{n-1}}$. To do so, each infoset for variant G_n has to be included in exactly one infoset with G_{n-1}, i.e., for all $n > 0$, for all $I \in \mathcal{I}_i^{G_n}$, there exists a unique $I' \in \mathcal{I}_i^{G_{n-1}}$ such that $I \subseteq I'$. This can be easily fulfilled by hypothetically including $\mathcal{H}^{G_n}$ in $\mathcal{I}^{G_{n-1}}$ as abstraction.

Because CFR with a sequence of variants, $(G_1, G_2, \ldots)$, is enhanced by the initialization using the former results in Step 3, we call our method a curriculum learning. A primary advantage of the ordered approach is in iterative improvement. Usually, we cannot expect how well a strategy learned for G_i behaves in the original game before any enhancement. Therefore, it is effective to start with the smallest variant, G_1, gradually improve the strategy along with a larger G_n, and stop once a sufficient variant is obtained.

4.1 Application to Cheat

We explain an example of our method when applying it to Cheat. By analyzing the game rule, we can see that to win the game, we want to not only keep as few cards as we can in our hand, but also win more challenges. On the basis of this thought, we bring "Health Points" (HPs) into this game.

In the original game, there is no restriction on how many times a player can lose challenges as long as no one wins the game. Now suppose each player has n HPs, which means they only have n chances to lose a challenge. More specifically, when a player's HPs becomes 0, they lose the entire game (even if their opponent has not discarded all their cards). Notice that the original winning condition still works but 0-HP losing condition has the higher priority (i.e.: if a player discards all the cards before any player's HP becomes 0, they win). We call Cheat with n HPs, Cheat-n. By limiting HPs, we created a technique of Ordered Abstraction. We propose to compute a smaller and easier version of the game, solve the game, and map the strategies into a larger game, i.e. sequentially solve Cheat-k, from Cheat-1 (the smallest variant), Cheat-2, ..., to obtain the strategy for the original Cheat, Cheat-∞.

When evaluating the playing performance of a strategy trained with Cheat-n in Cheat-n' where $n < n'$, an agent may be faced with an unknown situation, i.e., an infoset with n'' HPs where $n < n'' \leq n'$. In such cases, we use the strategy learned in case $n'' = n$, so we argue that our method is a type of abstraction.

5   Experiments

5.1 Experimental Setups

All experiments were conducted in a simplified version of Cheat, called *Mini-Cheat*. In Mini-Cheat, we use cards of three ranks and two cards for each rank, i.e. six cards in total. There are two players in the game and we deal two cards to each player to eliminate the possibility of perfect information. Although only a subset of cards is used in Mini-Cheat, it inherits an important property of infinite game length with repetition from the original.

In the following, "Cheat-n" refers to Mini-Cheat with n HPs for each player, unless stated otherwise. Moreover, we found that the average number of challenges in one game without any restriction is about 3.5, so we start with Cheat-3; a simple but still strategically complex version of the game.

To evaluate how our agents perform in different environments under various ways of training, we built two testing bots: Random and Heuristic. In both the *Discard* and *Challenge* phases, the Random bot chooses its action from all legal actions randomly with equal probability.

The Heuristic bot was built on the basis of human knowledge. It memorizes all the cards that it observed, keeps records of their locations, and makes decisions on the basis of its memory in a conservative way. In the *Discard* phase, it always discards honestly with the current rank as long as possible, or discards a random card among its private cards except for ones needed in its next turn. In the

Challenge phase, it challenges with probability 0% or 100% if the opponent's claim is consistent or not with its memory. If it cannot determine the consistency, it challenges with probability 50%. When it chooses not to challenge in this case, it believes the opponent and update its memory in accordance with the claim.

At the beginning of a game, any player cannot infer the opponent cards because a random subset of the cards are dealt but once the Heuristic bot identifies all the cards, then the bot plays perfectly. Please note that even though this does not happen, the Heuristic bot is still quite strong and more accurate in the *Challenge* phase.

Table 1. Time and Space Costs of four agents

	Memoryless		HP-Aware		History-Aware		Baseline	
	Time[1]	Space[2]	Time	Space	Time	Space	Time	Space
Cheat-3	60	292	52	1331	171	10689	224	11433
Cheat-4	860	292	1057	2557	5514	113331	5054	120512
Cheat-5	17887	292	19034	4257	18371	1209336	20315	1523174

[1] Time is in seconds.

[2] Space is represented in the number of infosets.

We tested the original CS-CFR and three variants of it. The difference between them is what they save in their infosets. All of them include the information about the setting of the game environment, the numbers of cards in the pile and in each player's hand, the card(s) in their hand, and the current rank. In addition, (1) Baseline (B) agent (original CS-CFR) includes both game history and current remaining HPs (i.e. all information it can obtain during the game); (2) History-Aware (HA) agent includes the game history in its infosets but not the remaining HPs; (3) HP-Aware (HPA) agent is aware of the HP information but not the game history; (4) Memoryless (M) agent does not include either the game history or HP information in its infosets. Similar to the naming of the game environments, we call Memoryless agents trained in Cheat-n, Memoryless-n (M-n). The same is applied for all the other agents.

5.2 Results

We compare all the agents in three aspects: time consumption and storage space costs, winning rates against testing bots in different game environments, and the effect of an agent's results on other's training processes (curriculum learning).

Table 1 shows the cost of four agents after the first 100 training iterations of CS-CFR. We can see that the time costs all increase exponentially as the game becomes more complex. The numbers of infosets of M agents stay constant. The growth of numbers of infosets of the HPA agent is exponential in powers of about two while that of the HA and B agents is almost 10. It is interesting that although HPA agent costs more in space, it needs less time than M agent. The same thing happens between B and HA agent.

To evaluate the learning efficiency and performance strength, we use the winning rate of the Heuristic bot against the Random bot as our baseline, which is approximately 82% (slightly varies in different game environments). The baseline will be represented in the red line in the following graphs.

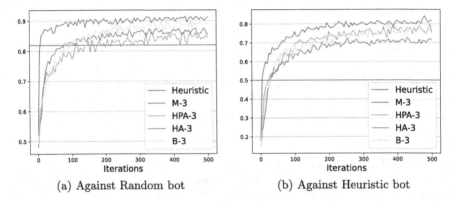

(a) Against Random bot   (b) Against Heuristic bot

Fig. 1. Winning rates of four variants against two testing bots: x-axis is the number of training iterations. The red line represents our baseline (Heuristic bot). (Color figure online)

We first test our agents against two testing bots every five training iterations; 500 iterations in total. Figures 1(a) and 1(b) reveal the winning rate trends of four agents in Cheat-3 against the Random and Heuristic bots, respectively. The x- and y-axes are the number of training iterations and winning rate, respectively. We notice that after 500 iterations, most agents become strong enough to exceed the baseline, and in particular, Memoryless-3 reached more than 90%. Agents even beat the Heuristic bot with winning rates over 70% while Memoryless-3 reached 80%. The reason that the Memoryless agent performs better is that the number of its infosets is the smallest, so we update its infosets more times than others during the same number of iterations, which results in a better strategy. We also notice that most agents have a much steeper learning efficiency at the beginning of the training and become steadier in later iterations.

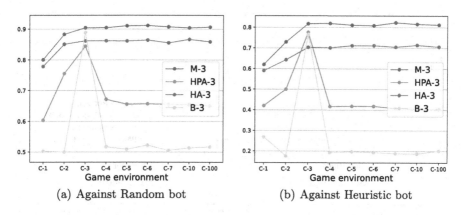

(a) Against Random bot   (b) Against Heuristic bot

Fig. 2. Generalization ability of four agents in different variants of Cheat: Winning rates of agents against Random and Heuristic bots in different game environments

Figure 2 demonstrates how agents perform in different game environments, from Cheat-2 to Cheat-100 (represented as C-2 to C-100). When the agents are in a different game (in particular, a larger game), there are game states (infosets) they never encounter in the training, and in such a case, they follow the strategy in the nearest (closest HPs information or history) infoset. We can see that the Memoryless and HA agents can perform better in the games that have larger numbers of HPs, while the HP-Aware is less strong and General agent only excels in the game environment that it was trained in. Notice that in Mini-Cheat-100, none of the players loses all their HPs, and the game always ends because someone discards all their cards. As a result, Mini-Cheat-100 is empirically the same as Mini-Cheat-∞.

We then test how the strategy profiles agents gained in smaller games affect the training in larger games. Instead of starting from scratch where the initial strategy and regret of each infoset are zero, we use the infosets of the agent trained in Cheat-n to initialize the training in Cheat-n' where $n' > n$. In Fig. 3, lighter lines represent agents trained from nothing while darker lines represent agents trained in Cheat-4 on the basis of the infosets data from Cheat-3. For example, the darker blue line in Fig. 3(a) is the winning rate of Memoryless-4 using Memoryless-3's final strategy profile at the beginning of the training.

From Fig. 3 we can see that abstractions provided by Ordered Abstraction training with a smaller game serves as a good approximation of that with a larger one for M, HA, and B agents since the darker lines start at higher places and are always higher than the lighter ones. However, it is less useful for HPA agents because the trends of lines of one type are almost the same.

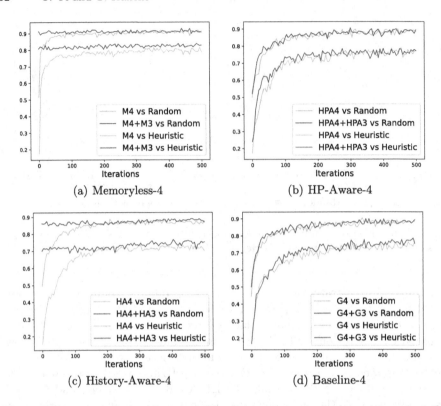

Fig. 3. Effect on curriculum learning: Blue lines represent winning rates against the Random bot; Green lines represent winning rates against the Heuristic bot. Lighter lines show training from scratch while darker lines show training based on former data. (Color figure online)

6   Conclusion

In this paper, we introduce Ordered Abstraction, an abstraction of limiting game lengths effectively in imperfect information games with a large or possibly infinite game length to make the training feasible. The idea is to make a variant where the game is forced to terminate in a finite number of steps. Also, by relaxing the condition of forced termination, we designed a curriculum learning with a series of variants from the most abstracted variant toward the original game.

Specifically in Cheat, we included a new term called "Health Points" that limits the number of challenges a player can lose in one game. With this method, we first designed smaller variants of Cheat so that training of Chance-sampled CFR agents becomes feasible. We then showed how we trained CFR agents and evaluated their performance against two testing bots. Moreover, we also demonstrated that we can utilize strategy profiles obtained in smaller games in the training of larger ones and the experiments show that there is an increase in the learning efficiency of specific agents.

For future work, we are also interested in including other abstraction techniques that can be independently used with ours to further improve learning efficiency to tackle the original Cheat between two or more players. A theoretical foundation and the ability to generalize to other games would also be an interesting line of further research.

References

1. Brown, N., Sandholm, T.: Simultaneous abstraction and equilibrium finding in games. In: Twenty-Fourth International Joint Conference on Artificial Intelligence (2015)
2. Brown, N., Sandholm, T.: Strategy-based warm starting for regret minimization in games. In: Thirtieth AAAI Conference on Artificial Intelligence (2016)
3. Brown, N., Sandholm, T.: Superhuman AI for heads-up no-limit poker: libratus beats top professionals. Science **359**(6374), 418–424 (2018)
4. Brown, N., Sandholm, T.: Superhuman AI for multiplayer poker. Science. **365**, 885–890 (2019)
5. Myerson, R.B.: Game Theory: Analysis of Conflict. Harvard University Press, Cambridge (1997)
6. Neller, T.W., Hnath, S.: Approximating optimal Dudo play with fixed-strategy iteration counterfactual regret minimization. In: van den Herik, H.J., Plaat, A. (eds.) Advances in Computer Games, pp. 170–183. Springer, Heidelberg (2012)
7. Zinkevich, M., Johanson, M., Bowling, M., Piccione, C.: Regret minimization in games with incomplete information. In: Advances in Neural Information Processing Systems, pp. 1729–1736 (2008)

Deep Reinforcement Learning for Morpion Solitaire

Boris Doux, Benjamin Negrevergne, and Tristan Cazenave[✉]

LAMSADE, Université Paris-Dauphine, PSL, CNRS, Paris, France
Boris.Doux@dauphine.psl.eu, Tristan.Cazenave@lamsade.dauphine.fr

Abstract. The efficiency of Monte-Carlo based algorithms heavily relies on a random search heuristic, which is often hand-crafted using domain knowledge. To improve the generality of these approaches, new algorithms such as Nested Rollout Policy Adaptation (NRPA), have replaced the hand crafted heuristic with one that is trained *online*, using data collected during the search. Despite the limited expressiveness of the policy model, NRPA is able to outperform traditional Monte-Carlo algorithms (i.e. without learning) on various games including Morpion Solitaire. In this paper, we combine Monte-Carlo search with a more expressive, non-linear policy model, based on a neural network trained beforehand. We then demonstrate how to use this network in order to obtain state-of-the-art results with this new technique on the game of Morpion Solitaire. We also use Neural-NRPA as an expert to train a model with Expert Iteration.

1 Introduction

Monte-Carlo search algorithms can discover good solutions for complex combinatorial optimization problems by running a large number of simulations. Internally, the simulations are used to evaluate each alternative branching decision, and the search algorithm successively commits to the best branching decision until a terminal state is reached. Thus, one can see simulations as a tool to turn uninformed (random) search policies into well informed ones, at the cost of computational power. Building on this observation, *Nested Monte Carlo Search* (NMCS) further improves the technique by running recursive (a.k.a. nested) simulations. At the lowest recursive level, the simulations are driven by a simple random search policy. At higher recursive levels, the simulations are driven by a search policy that is based on the simulations of the recursive level below. Nesting simulations greatly improve the quality of the solutions discovered, however it is generally impossible to run NMCS with more than 5 or 6 levels of recursion, due to the prohibitive cost of recursive simulations.

To further improve the quality of the results, it is often desirable to replace the purely random search policy with a hand crafted search heuristic, but building such heuristic is time consuming and requires expert knowledge which is difficult to encode in the search heuristic. To overcome this limitation and facilitate the adaptation of Monte-Carlo search to new problems, *Nested Rollout Policy Adaptation* [10] replaces the recursive policies, with a simple policy model that is *learned*, using data collected during the search. Thanks to this simple principle, NRPA is now the state of the art on different problems such as vehicle routing problems, network traffic engineering or RNA design as well as the game of *Morpion Solitaire* which became a testbed for several Monte-Carlo based algorithms such as NRPA and NMCS.

However, despite the success of learned policies, and a number of recent studies on the topic, the last major record break on Morpion Solitaire dates back from 2011. (Rosin obtained 82 on the 5D variant with a week long execution of NRPA).

Recently [2] has managed to rediscover the best score with optimized play-outs, but despite many tries was unable to break the record. The recent success of AlphaGo/AlphaZero [12–14] suggests that combining Monte-Carlo search together with a neural network based heuristic can lead to important improvements. AlphaZero like Deep Reinforcement Learning has been tried for Morpion Solitaire with PUCT [15].

In this paper, we look into learning an expressive policy model for the Morpion Solitaire that is based on a deep neural network, and we use it to drive simulations at low computational cost. We then conduct thorough experiments to understand the behaviour of new and existing approaches, and to assess the quality of our policy models. Then we reintroduce this neural network based policy inside NMCS. We are able to obtain a policy which is almost as good as state-of-the-art NRPA algorithm with 3 nested levels, for a 2–3 times reduction of computational time. Finally, we experiment using self-play with a second approach based on Expert Iteration (Exit) with various experts. Our approach is able to learn a policy from scratch and outperforms previous work on selfplay in Morpion Solitaire by 6 points.

The rest of this paper is organized as follows: the second section describes related work on Monte Carlo Search. The third section explains search with a learned model. The fourth section shows how to combine neural networks and Monte Carlo Search. The fifth section shows how to apply Deep Reinforcement Learning using Neural NMCS and Neural NRPA. The sixth section outlines future work.

2   Preliminaries on Monte-Carlo Search for Game Playing

Policies: A policy is a probability distribution p over a set of moves $\mathcal{M}$ that is conditioned on the current game state $s \in S$. For example, we often consider the uniform policy p_0, which assigns equal probability to all the moves that are legal in state s. I.e. $p_0(m|s) = \frac{1}{|M_s|}$.

In this paper, we also consider policies probability distributions p_W which are parameterized with a set of weights W. There is one real valued weight for each possible move, i.e. $W = w_{m_1}, \ldots, w_{m_{|\mathcal{M}|}}$, and the probability $p_W(m|s)$ is defined as follows:

$$p_W(m|s) = \frac{e^{w_m}}{\sum_{p \in M_s} e^{w_p}}$$

The softmax function enables to calculate the gradient for all the possible weights associated to the possible moves of a state and to learn a policy in NRPA using gradient descent.

Finally, we also consider more complex policies π_θ in which the probability of each move depends on a function of the state, represented using a neural network. Let $f_\theta : S \rightarrow \mathbb{R}^{|M|}$ be a neural network parameterized with θ, we can then define policy π_θ as follows:

$$\pi_\theta(m|s) = \frac{e^{(f_\theta(s))_m}}{\sum_{p \in M_s} e^{(f_\theta(s))_p}}$$

2.1 NMCS and NRPA

As most Monte-Carlo based algorithms, *Nested Monte Carlo Search* (NMCS) and *Nested Rollout Policy Adaptation* (NRPA) both generate a large number of random sequences of moves. The best sequence according to the scoring function is then returned as a solution to the problem. The quality of the final best sequence directly depends on the quality of the intermediate random sequences generated during the search, and thus on the random policy. Therefore NMCS and NRPA have introduced new techniques to improve the quality of the policy throughout the execution of the algorithm.

NMCS and NRPA are both recursive algorithms, and at the lowest recursive level, the generation of random sequences is done using playouts parameterized with a simple stochastic policy. If the user has access to background knowledge, it can be captured by using a non-uniform policy (typically by manually adjusting the weights W of a parameterized policy p_W). Otherwise, the uniform policy p_0 is used.

In NMCS, the policy remains the same throughout the execution of the algorithm. However, the policy is combined with a tree search to improve the quality over a simple random sequence generator. At every step, each possible move is evaluated by completing the partial solution into a complete one using moves sampled from the policy. Whichever intermediate move has led to the best completed sequence, is selected and added to the current sequence. The same procedure is repeated to choose the following move, until the sequence has reached a terminal state.

A major difference between NMCS and NRPA, is the fact that NRPA uses a stochastic policy that is *learned* during the search. At the beginning of the algorithm, the policy is initialized uniformly and later improved using gradient descent based the best sequence discovered so far. The policy weights are updated using gradient descent steps to increase the likelihood of the current best sequence under the current policy.

Finally, both algorithms are nested, meaning that at the lowest recursive level, weak random policies are used to sample a large number of low quality sequences, and produce a search policy of intermediate quality. At the recursive level above, this policy is used to produce sequences of high quality. This procedure is applied recursively. In both algorithms the recursive level (denoted *level*) is a crucial parameter. Increasing *level* increases the quality of the final solution at the cost of more CPU time. In practice it is generally set to 4 or 5 recursive level depending on the time budget and the computational resources available.

2.2   Playing Morpion Solitaire with Monte Carlo Search

The game of Morpion Solitaire. Morpion Solitaire is a single player board game. The initial board state is shown in Fig. 1 and a move consists of drawing a circle on an empty intersection, and drawing a line out of five neighboring circles including the new one. A game is over when the player runs out of moves, and the goal of the game is to play as many moves as possible. The final score is simply the number of moves that have been played. There are two versions of the game called 5T (T for *touching*) and 5D (D for *disjoint*). In 5T two lines having the same direction can share a common circle, whereas in 5D they cannot.

The best human score for 5T is 170 moves and it has been discovered by Charles-Henri Bruneau who held this record for 34 years until he was beaten by an algorithm based on Monte-Carlo search. The current best score is 82 for 5D and 178 for 5T. Both records were established in August 2011 by Chris Rosin with an algorithm combining nested Monte-Carlo search and a playout policy learning (NRPA, [3, 10]).

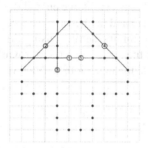

Fig. 1. Move 1, 2, 3 and 4 are legal for 5D and 5T. Move 5 is legal for 5T only

Modeling Morpion Solitaire as a Monte-Carlo Search Problem. Any game state is fully determined by the set of (oriented) segments connecting the circles. Thus, the initial game state s_0 is the empty set, and performing a move consists of adding a segment to the set of segments representing the current state. Each segment (or move) is determined by a 2D coordinates representing the starting point of the segment, and one direction among the 4 possible directions: left to right, top to bottom, top-left to bottom-right, and top-right to bottom-left. The game is over when the player reaches a terminal state i.e. a state s such that $M_s = \emptyset$.

Although the order in which the moves are added does not influence the final game state, (i.e. for any sequence of moves X and any permutation X' of X, we have $state(X) = state(X')$), it is generally difficult to compute the subset of moves that can be added without breaking the rules. Therefore the moves are drawn sequentially such that every intermediate state is also a legal state.

3 Imitating NRPA

In this section, we first focus on training a policy model that can be used to select good moves, without having to simulate a large number of games. We recall that a policy model is a conditional probability distribution $\pi_\theta(m|s)$ where s is a game state from the set of all possible game states S, and m is a move from the set of all possible move $\mathcal{M}$.

To obtain a good policy, we first train our policy model to learn to reproduce the sequences found by NRPA. The policy model is represented by a neural network, and is trained to predict the next NRPA move, given a description of the current game state. Each supervised example is a particular game state, labeled with the move that was chosen by NRPA during a previous run. (Note that since NRPA is a stochastic algorithm, identical game states may appear several times in the dataset, labeled with different moves.)

To successfully reproduce sequences found by NRPA, we need 1: a game state representation that contains the adequate features to accurately predict the next move by NRPA, and 2: a policy model that is expressive enough to capture the complex relation that exists between the game state and the best move selected by NRPA. In this section, we design and evaluate several training settings using different game state representations and different models. We then discuss the performance of these settings by using two criteria: the ability to mimic the behaviour of NRPA, and the quality of a play (i.e. the game score).

3.1 Game State Representation

Although the game state is fully determined by the set of segments (as discussed in Sect. 2), this representation does not favor learning, and generalization over different but similar states. In this section, we discuss a better state representation, that explicitly captures important features and makes it possible to predict the behavior of NRPA, without having to run the costly simulations.

In all our models, the board is represented by five 30×30 binary valued matrices, which are large enough to capture any record holding boards. The first matrix is used to represent the occupied places (i.e. the circles in Fig. 1) which are not directly available nor easy to compute from a board state represented as a set of segments. If the place i, j is occupied on a board, the corresponding value in the matrix is set to 1, and 0 otherwise.

Because this matrix alone does not fully determine the game state, the four extra binary valued matrices are used to represent the connecting segments, one matrix for each possible direction respectively: left to right, top to bottom, top-left to bottom-right, and top-right to bottom-left. A one in the first matrix (left to right) at position i, j signifies that there is a segment between the place i, j and the place $i + 5, j$ on the board. A one in the second matrix (top to bottom) at position i, j signifies that there is a line between position i, j and position $i, j + 5$ on the board and so on for each matrices. Every time a new place is occupied (i.e. the player makes a move) we set one boolean value in the first matrix, and one boolean value from one of the 4 remaining matrices.

In addition to the board representation, we extend the state representation with 4 extra matrices which are meant to represent all the possible moves for the next move (one matrix for each possible direction). We call this first representation **R1**.

To further improve temporal consistency of the policy model, we extend the first state representation with 8 extra matrices, to represent the 8 previous moves. We call this second representation **R2**.

3.2 Neural Network Architecture

We consider two neural network architectures. The first one is a fully convolutional neural network with 4 convolutional layers. The first 3 layers have 32 filters with 3×3 kernels, and the last convolutional layer has 4 filters with 1×1 kernels to match the output. The output is a vector of dimension $n^2 d$ where n is the dimension of the board and d the number of directions to represent all possible moves (in all our experiments use n = 30 and d = 4).

The second architecture is a residual neural network [8] with 4 convolutional layers with the same type and number of filters as the first architecture, and the same input/output definition.

We found the use of a fully convolutional model more effective than the policy heads used in AlphaGo and Alpha Zero which contain fully connected layers. A fully convolutional head is similar to the policy heads of Polygames [7] and Golois [6].

3.3 Training Data and Training Procedure

We train the policy models using data generated with NRPA. Each example in the training set is a game state representation labelled with one move played by NRPA in this game state. To improve the quality of the training data, we can select only the moves from the NRPA games that scored well, however it is important to remark that there is a trade-off between the quality of the moves, and the diversity of the training data (a.k.a. the exploration vs. exploitation trade-off). To observe this phenomenon, we selected 10 000 games (800.000 game states) generated with NRPA that scored 80 or above (Fig. 2 first plot), and 10 000 games (around 800.000 game states) generated with NRPA that scored between 70 and 82 (Fig. 2 second plot). As we can see in the first plot, game states dramatically lack of diversity.

Based on this observation and other empirical analysis, we used NRPA to generate a large number of games, and selected 9141 games scoring between 70 and 82 for a total of 694 716 training examples (couples: game state, move). We use this data to train the neural networks models described above, using the two representations R1 and R2. We used a decaying learning rate starting at 0.01 and divided it by 10 every 40 epochs.

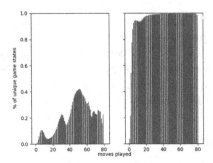

Fig. 2. NRPA data diversity

3.4   Model Performance (Without Search)

We now compare the different policy models using two metrics. First we consider the test accuracy, that is how well the policy models are able to predict NRPA moves. Then we estimate the average and the maximum score obtained with a playing strategy which sample moves from the different policy models. The mean and the average are computed over 100 000 games.

Train/Test accuracy. We first compare the two neural network architectures and the two game state representations that we have described in the previous section. The values of the loss functions during the training procedure for each architecture are shown in Fig. 3, and a comparison of the accuracy achieved by each architecture and each state representation is shown in Fig. 4. (We only show the comparison of the state representation using the Resnet architecture since it performs best.)

We first consider, the initial model with the game state representation R1 and the BasicCNN neural network architecture shown in Fig. 3 (left). We observe that the training loss quickly reaches its lowest value, and that an important difference between the training and the testing loss remains. Unsurprisingly, this results in a poor model accuracy of 45.5% on the test set (as seen in Fig. 1). Furthermore, this peculiar behaviour is not impacted by the use of a larger, more expressive neural network architecture such as the Resnet or by any more sophisticated training procedure.

To explain this behaviour, we recall that 1) NRPA is not deterministic, 2) the policy in NRPA is trained in a stochastic way and may vary significantly from one game to another. Non determinism leads to presence of a large number of identical examples labelled differently in the train and test set, which induces an incompressible Bayes risk, that cannot be removed, by increasing the expressivity of the model, or by improving the training procedure.

However, the behaviour is remarkably different on the second representation R2 which includes the previous moves in addition to current game state. This may be surprising, since with an unbiased algorithm, the best move only depends on the current state, and should not depend on the previous actions performed by the player. However, NRPA is biased by the learned policy, which differs from one game to another. The previous moves thus informs the neural network on the current policy, and the particular strategy that is being played, and ultimately reduces Bayes's risk. As a result, the neural network is able to better fit the training set (and benefits from additional epochs), the final loss is lower, the generalization gap is reduced, and the final accuracy reaches 70%.

This suggests that unlike the first two models based on R1, last model based on R2 is able to capture not just one strategy but several good strategies that were discovered by NRPA during the 9141 selected games.

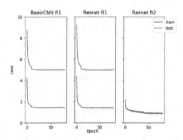

Fig. 3. Loss evolution during training

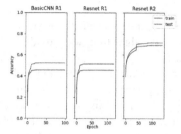

Fig. 4. Accuracy evolution during training

Table 1. Accuracy for each tested configuration

Epoch	BasicCNN R1	Resnet R1	Resnet R2
1	25%	27%	47%
40	45.5 %	45.3 %	67%
80	45.5 %	45.5 %	68.9%

Score. To evaluate the quality of the policy models as players, we sample sequence of moves from each policy model and observe the score of the final state. The distribution of the scores across 100 000 sequences generated from each policy model is shown in Fig. 5.

In both plots, we have a high probability of reaching a score between 57 and 62. However, the second model based on state representation R2 demonstrates better results in the early games, and there are fewer games that score less than 50 points. We believe that the second game state representation, which includes the previous moves, is able to achieve better temporal consistency and avoid simple mistakes which may be the consequence of mixing several NRPA strategies from the training set. The model based on R2 also exhibits the highest average score, and maximum score than the model based or R1.

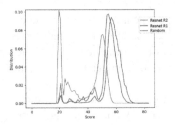

Fig. 5. Score distribution of Resnet R1 and Resnet R2

To accurately evaluate the quality of the models and to compare it with the original NRPA algorithm, we provide more precise score statistics which are available in Table 2.

In this table, *Uniform* is the performance of the uniform policy model p_0, the next 3 are the performance of NRPA with increasing level recursions, and the last 3 are our models, described in the previous sections. The statistics for the Uniform policy and our models, are averaged over 100 000 games. However, generating NRPA games is computationally intensive so the statistics for NRPA(1), NRPA(2) and NRPA(3) are computed over 100 000, 10 000 and 400 games respectively where number between brackets refers to the number of recursive levels

We can see that the two neural network models based on R1 (without the previous moves) offer a little improvement over the baseline, but are outperformed by NRPA(1). However the neural network based on R2 performs significantly better than the baseline, (mean and max), and achieves better maximum scores than NRPA(1) and NRPA(2), without having to run a large number of rollouts.

Table 2. Results of our approaches compared to state of the art algorithm.

	mean	max	$\sigma/\sqrt{n}$
Uniform	39.1	61	0.059
NRPA(1)	58.5	66	0.014
NRPA(2)	65.9	72	0.024
NRPA(3)	68.2	78	0.119
BasicCNN R1	41.7	60	0.024
Resnet R1	44.0	58	0.018
Resnet R2	50.5	74	0.032

3.5 Combining MC Search Algorithms with a Neural Based Search Procedure

We now have a neural network that can act as an informed search heuristic comparable to a NRPA of level 2–3. To further improve the quality of the solutions, we incorporate the newly trained policy model inside existing search algorithms, in place of the random heuristic.

Table 3 summarizes results achieved by the different policy models. Nested(1) where number between brackets refers to the number of recursive levels, outperforms Resnet by 18,3 points in average and by 5 points for the maximum score. In this setup, our approach outperforms NRPA(2) in mean and maximum and perform very close to NRPA(3).

Table 3. Comparison between different search algorithms

	mean	max	avg. game time
NRPA(3)	68.2	78	16:40
Nested(1) + Resnet R2	**68.8**	**79**	**6:26**
Resnet R2	50.5	74	0:01

4 Self Play with *Exit*

In the previous section, we were able to obtain a playing strategy by training a neural network with game data generated by NRPA. Although the resulting strategy is good and computationally efficient, this technique remains entirely supervised by NRPA, and thus it is unlikely performing better than NRPA itself.

In this Section, we explore *self-play* and learn a new policy from scratch using an approach based on Exit [1]. In contrast with the previous approach in which the neural network is only used to store and generalize past experiences acquired through supervision, in Exit the expert is also based on a neural network and can be improved as we discover new good moves. This allows the expert to learn from scratch, and improve beyond the current best known strategy. (See [1] for details.)

Exit has been used in the notorious Alpha Zero [13] and Wang et al. [15] applied it for Morpion Solitaire. However, our approach is different since it does not use PUCT as the search algorithm. Instead we use an expert based on NRPA which is state-of-the art in Morpion Solitaire. Although using NRPA poses a number of challenges, we are able to outperform state-of-the-art in the self play setting by a significant margin.

Speeding up NeuralNRPA. The main challenge that is to overcome if we want to use NeuralNRPA as an expert is the computational cost. Despite the improvement discussed in the previous Section training a policy from scratch using NeuralNRPA remains prohibitive.

In the previous approach, we make a forward pass at each step, which induces a significant computational cost. In the Morpion Solitaire, moves are often commutative, meaning that playing move a, then b leads to the same state than playing move b, then a. We can exploit this property and make a single forward pass for an entire game (including the many rollouts). This results in a small reduction of the average score, but a dramatic reduction of computational cost.

Training Setting. In our experiments, at each iteration we generated 10.000 boards with the learner. We train our model with a learning rate of 5.10^{-4} and 20 epochs.

We tested Exit using 3 different configurations: **NeuralNMCS(0)** is Exit with NMCS lvl 0 as expert which means the expert use the best sequence out of x rollouts played by the neural network. **NeuralNMCS(1)** is Exit with NMCS level 1 as expert, **NeuralNMCS(1)c** the expert is a early stopped version of NMCS where instead of calling NMCS after each move played, we stop the algorithm after the end of the first call from the highest nested level end and use the best sequence found as the label instead of choosing only one move after one call and repeat until the end of the sequence. **NeuralNRPA(1)** and **NeuralNRPA(2)** are GNRPA [5] of level 1 and 2 with the bias given by the policy output by the neural network. GNRPA had a bias to NRPA action's weight leading to a bootstraped NRPA into a specific direction.

Table 4 gives mean and max scores of neural networks trained by the different approaches. All of the approaches have been running for 180 h. NeuralNRPA(2) A, NeuralNMCS(1)c A, NeuralNMCS(1) A and NeuralNMCS(0) A are the best approaches among all tested parameters.

The Fig. 6 displays the evolution of the maximum score evolution on 100 rollouts with the four best approaches of each type. NeuralNMCS(0) A is the fastest reaching 70 but it gets stuck quickly. NeuralNMCS(1) A shows poor exploration due to a low number of rollouts but it is also very slow regarding the number of its rollout parameter. NeuralNMCS(1)c A and NeuralNRPA(2) A are slower then NeuralNMCS(0) A but ended up outperforming it. NeuralNMCS(1)c A is a bit faster than NeuralNRPA(2) A at the beginning but NeuralNRPA(2) A gets the highest score at the end.

Table 4. Comparison of approaches

Approach	Temperature	Rollouts	NN mean score	NN best score	Best score in (Hours)
NeuralNMCS(0)	0.2	1	50.66	64	6
NeuralNMCS(0)	0.2	100	63.76	68	72
NeuralNMCS(0) A	0.4	100	54.33	70	51
NeuralNMCS(1) A	0.2	1	56.1	66	50
NeuralNMCS(1)c	0.2	1	61.42	68	63
NeuralNMCS(1)c A	0.2	10	64.38	72	134
NeuralNMCS(1)c	0.4	10	53.23	67	111
NeuralNRPA(1)	0.2	100	47.4	63	3
NeuralNRPA(2)	0.2	10	54.9	71	93
NeuralNRPA(2) A	**0.2**	**20**	**57.28**	**73**	**180**
NeuralNRPA(2)	0.2	40	53.78	70	151

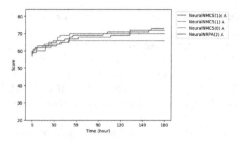

Fig. 6. Best approaches max score evolution

5 Conclusion

We have shown that it is possible to learn an exploratory policy for Morpion Solitaire from a set of states with score ranging from high average scores (70) to highest known scores (82) using a neural network. We also integrated this neural network in a Nested Monte Carlo search and showed it improves when sampling from its moves distribution reaching scores 3 moves away from the highest known score. We have also trained a network with an original version of Expert Iteration using Neural NRPA and Neural NMCS and found that Neural NRPA is the best expert performing 6 points higher than the reinforcement learning approach in [15].

In computer Go and more generally in board games the neural networks usually have more than one head. They have at least a policy head and a value head. The policy head is evaluated with the accuracy of predicting the moves of the games and the value head is evaluated with the Mean Squared Error (MSE) on the predictions of the outcomes of the games. The current state of the art for such networks is to use residual networks [4,13,14]. The architectures used for neural networks in supervised learning and Deep Reinforcement Learning in games can greatly change the performances of the associated game playing programs. For example residual networks gave AlphaGo Zero a 600 ELO gain in playing strength compared to standard convolutional neural networks. Mobile Networks [9,11] are commonly used in computer vision to classify images. They obtain high accuracy for standard computer vision datasets while keeping the number of parameters lower than other neural networks architectures. For board games and in particular for Computer Go it was shown recently that Mobile Networks have a better accuracy than residual networks [6].

We plan to try different architectures for Morpion Solitaire neural networks and compare their performances.

References

1. Anthony, T., Tian, Z., Barber, D.: Thinking fast and slow with deep learning and tree search. In: Advances in Neural Information Processing Systems, pp. 5360–5370 (2017)
2. Buzer, L., Cazenave, T.: Playout optimization for Monte Carlo search algorithms. Application to Morpion Solitaire. In: IEEE Conference on Games (2021)
3. Cazenave, T.: Nested Monte-Carlo search. In: Boutilier, C. (ed.) IJCAI, pp. 456–461 (2009)

4. Cazenave, T.: Residual networks for computer go. IEEE Trans. Games **10**(1), 107–110 (2018)
5. Cazenave, T.: Generalized nested rollout policy adaptation. arXiv preprint arXiv:2003.10024 (2020)
6. Cazenave, T.: Mobile networks for computer go. IEEE Tans. Games (2020)
7. Cazenave, T., et al.: Polygames: improved zero learning. ICGA J. **42**(4), 244–255 (2020)
8. He, K., Zhang, X., Ren, S., Sun, J.: Deep residual learning for image recognition, pp. 770–778 (2016)
9. Howard, A.G., et al.: MobileNets: efficient convolutional neural networks for mobile vision applications. arXiv preprint arXiv:1704.04861 (2017)
10. Rosin, C.D.: Nested rollout policy adaptation for Monte Carlo Tree search. In: IJCAI, pp. 649–654 (2011)
11. Sandler, M., Howard, A., Zhu, M., Zhmoginov, A., Chen, L.C.: Mobilenetv 2: inverted residuals and linear bottlenecks. In: Proceedings of the IEEE Conference on Computer Vision and Pattern Recognition, pp. 4510–4520 (2018)
12. Silver, D., et al.: Mastering the game of go with deep neural networks and tree search. Nature **529**(7587), 484–489 (2016)
13. Silver, D., et al.: A general reinforcement learning algorithm that masters chess, shogi, and go through self-play. Science **362**(6419), 1140–1144 (2018)
14. Silver, D., et al.: Mastering the game of go without human knowledge. Nature **550**(7676), 354–359 (2017)
15. Wang, H., Preuss, M., Emmerich, M., Plaat, A.: Tackling Morpion Solitaire with AlphaZero-like ranked reward reinforcement learning (2020)

Expert Iteration for Risk

Lucas Gnecco Heredia and Tristan Cazenave[✉]

LAMSADE, Université Paris-Dauphine, PSL, CNRS, Paris, France
Tristan.Cazenave@dauphine.psl.eu

Abstract. Risk is a complex strategy game that may be easier to understand for humans than chess but harder to deal with for computers. The main reasons are the stochastic nature of battles and the different decisions that must be coordinated within turns. Our goal is to create an artificial intelligence able to play the game without human knowledge using the *Expert Iteration* [1] framework. We use graph neural networks [13,15,22,30] to learn the policies for the different decisions and the value estimation. Experiments on a synthetic board show that with this framework the model can rapidly learn a good country drafting policy, while the main game phases remain a challenge.

1 Introduction

The game of Risk might be somehow simpler for humans compared to chess or Go. It has a much more complex game flow, but it requires less experience to be able to play at a decent level. For Risk, once the game rules and objectives are understood, human players can find out common sense strategies that work fairly well. Nevertheless. this classic game presents a lot of challenges when it comes to computer play.

To begin with, in Risk each turn consists on different phases that involve multiple decisions that should be coordinated. Moreover, the attack phase is stochastic because the result of the attack depends on a dice roll, introducing chance nodes to the game tree [18]. In terms of the number of players, it can be played with two to six players, so by nature it falls out of the usual two-player zero sum game category, opening the door to questions about coalitions and the impact of other players on the outcome of the game [19,27]. It has imperfect information because of the game cards that players can use to trade armies. Under traditional rules, trading cards will augment the number of armies for the next trade, making it less obvious to decide when to trade.

The present work aimed to create a Risk player able to learn *tabula rasa*, meaning without human knowledge and just using self-play. When we state *without human knowledge* we mean letting the player discover strategies by mere self-play and only from information that can be deduced directly from the board and game history. This can be summarized in the following points:

- Features used as input may be sophisticated but must be deducible from the information available to the player. This features should be as simple as possible, ideally just a property obtained directly from the board with some reasonable pre-processing (normalization, scaling, etc.)

© Springer Nature Switzerland AG 2022
C. Browne et al. (Eds.): ACG 2021, LNCS 13262, pp. 27–37, 2022.
https://doi.org/10.1007/978-3-031-11488-5_3

– We want to avoid hard coding strategies for particular board configurations. This also goes for the different game phases, meaning we would like to keep them *fundamentally* unchanged if possible. We believe a general thumb rule should be that the model used should be able to play another turn based, multiplayer game that contains chance nodes and multi-decision turns without incurring into *fundamental* changes.

The player and training methods follow the *Expert iteration* framework [1], closely related to *AlphaZero* [24]. For the policy and value neural networks we used graph neural networks [13, 15, 22, 30], given the fact that the Risk board can be naturally represented as a directed unweighted graph.

Section 2 covers the basic concepts needed to understand the whole approach. Section 3 is a small survey on previous attempts to create a Risk AI. Section 4 presents the design of the neural-MCTS player and implementation details. Section 5 explains the experimental setup including the modifications to the game rules made to simplify the problem. Results on a small map are shown and discussed to understand the possible future work directions. The code related to the project is available as a Github repository[1].

2   Preliminaries

2.1   Monte Carlo Tree Search

Monte Carlo Tree Search [9] is a search method that iteratively builds the search tree relying on random sampling to estimate the value of actions. The general algorithm consist on 4 phases: *selection, expansion, simulation* and *backpropagation*. The idea is that at each iteration the tree is traversed using a *tree policy* for choosing moves. When a leaf is found usually the tree is expanded by adding a new node. From there a *default policy* is used to play further moves (usually random) and end the episode. The results are then backpropagated through the traversed nodes [5].

The most known MCTS algorithm is probably Upper Confidence Tree (UCT) created in 2006 by Levente Kocsis and Csaba Szepesvári [14]. It makes use of the *UCB1* [3] bandit formula to choose moves along the tree descent. Rosin [21] later added prior knowledge to the formula to bias the initial choices towards moves suspected of being strong. This idea was called Predictor+UCB (PUCT) and it is the way in which the *AlphaZero* and *Expert Iteration* frameworks combine neural networks and MCTS. The idea is that neural networks can learn strong policies that can work as priors biasing the search towards promising nodes. They can also learn to evaluate game states which also improves the search process.

2.2   Expert Iteration

Expert Iteration [1, 2] is an effective algorithm to learn a policy given by a tree search algorithm like Monte Carlo Tree Search (MCTS). The idea is that a deep neural network

---

[1] https://github.com/lucasgneccoh/pyLux.

can approximate the policy provided by the MCTS and generalize it so that it can be used afterwards without performing the search, or it can be used to bias the search and get better results. At each iteration of the training process, the current neural network is used to guide the search of the MCTS and/or to evaluate game states at the leaves of the tree, resulting in a more efficient search and therefore a stronger policy. This new policy is learned again by the deep neural network and the process repeats.

3   Previous Work

The first Risk AI players to the best of our knowledge date back to 2005, where Wolf [28] tried to create a linear evaluation function. Apart from the value function, he programmed different plans that represent common sense strategies and allowed his player to play accordingly. A similar approach was developed in 2013 by Lütolf [17].

Another approach developed in 2005 was made by Olsson [11,20] who proposed a multi-agent player that placed one agent in every country plus a central agent in charge of communication and coordination. Each country agent works like a node in the map graph, passing and receiving information from its neighbors to evaluate its own state. Then every agent participates in a global, coordinated decision. It is worth highlighting the importance it gives to the graph structure of the board and the message passing between countries, concepts highly related to graph neural networks.

The first attempt that used MCTS for playing Risk was done in 2010 by Gibson, et al. [10] where they used UCT only on the initial country drafting phase of the game. They concluded that the drafting phase was key to having better chances of winning the game, and that their UCT drafting algorithm was able to make an existing AI player improve considerably.

In 2020, two contributions were made using neural networks. The one developed by Carr [6] uses temporal difference and graph convolutional networks (GCN) [13] to take information from the board and return an evaluation. This evaluation is used together with a breadth-first search to find the most promising end-turn state for the player. The possible end-turn states are enumerated by making the attack phase deterministic. Carr generates data using the available bots in the open-source game Lux Delux[2]: by Sillysoft. This games are used to train the evaluation function.

The key differences between Carr's approach and ours is that we wanted to learn using self-play only and that we wanted to model the actions in the game in a general way. This meant not using special techniques for complicated game phases like turning the attack phase deterministic by creating a look up table of possible outcomes [6], but relying only on the traditional tree search expanding each node by considering possible actions, whatever their nature. We also wanted to use MCTS instead of other tree search algorithms.

We would like to highlight that Carr's player proved to be good in the six player player game against strong enemies from Lux Delux, which is very impressive and shows that graph neural networks can be useful at extracting features and information from the game board. Again, our idea was to push a little further into the *tabula-rasa*

---

[2] https://sillysoft.net/lux/.

scheme and also to reduce the amount of hand crafted features. Unfortunately our approach was not able to reach this level of play. In Sect. 6 we will discuss on what could be done to improve the level of a player following our approach.

Coming back to Carr's approach, we think that one important fact that contributed to his player being successful was turning the attack phase deterministic. On our case, the value estimation of states and actions through sampling was particularly difficult and demanded a large number of simulations, so we think that having something like a look-up table with outcomes and probabilities could really speed up value estimation. In our approach we wanted to avoid using the probabilities of the outcomes in an attack so that the player could remain general and work in situations where these probabilities cannot be easily computed or may not be fixed.

The second player proposed in 2020 by Blomqvist [4] followed the *AlphaZero* algorithm. This is very similar to what we intend doing, but the network architecture consists only on fully connected linear layers. One valuable result is that even if the learned policies are not remarkably strong, they improve the MCTS when included as priors, allowing to conclude that they indeed bias the search towards interesting moves.

4 Player Design

4.1 General Player Design and Architecture

The first thing to notice about the game of Risk is that the board can be naturally represented as a graph. This immediately suggests the use of a Graph Neural Network (GNN) [22] instead of more traditional networks. Moreover, having seen the importance of convolutional layers on the development of the Go players [23], we decided to use Graph Convolutional Networks [13].

Following the same line of thought presented by Olsson [20], we considered countries as the fundamental blocks for any reading of the game board as most of the information needed is stored at a country level. We considered only basic information as input, such as the percentage of armies in the country from the board's total, the owner of the country, the continent it makes part of, the bonus this continent has, etc. Just as in Go [23], more complex features could be created such as if the country is on the border of the continent or not, if it has an ally/enemy neighbor. Once this features are computed, each country yields a tensor that is fed to deep residual convolutional layers [15,16] to create a hidden board representation. We were inspired by the increase in performance due to residual blocks and deeper models in the game of Go [7,25]. We used four deep GCN layers for the initial block and another four for each independent head.

One key idea we wanted to keep present in the network design was that for every action except for the card trade, a policy could be represented as a distribution over either the nodes or the edges of the graph. In our model, each head ends up with something similar to the input: one tensor for each node of the graph that can then be easily transformed into the desired policy. This makes the design flexible enough to adapt to any map and is similar to how the Go players were designed, reading as input a game board and giving as output a distribution that has the same shape [8,29]. Figure 1 shows a general diagram of the network design.

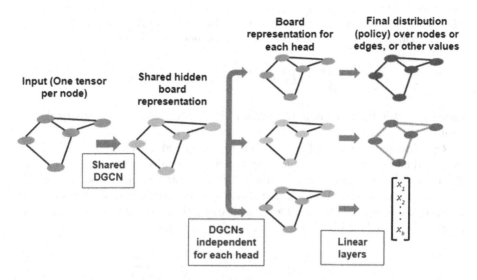

Fig. 1. General flow of the network. DGCN stands for Deep Graph Convolutional Network, which consist of layers of graph convolutions with normalization and a skip connection. They are implemented in PyTorch Geometric based on [15, 16]. The input consists on one tensor per node (country). The first DGCN will output again one tensor per node that we call hidden representation, and that is common to every head. This hidden representation is the input for each one of the heads, that will apply an additional DGCN with independent weights for every head. The output of each head is then passed through linear layers to obtain the desired shape, whether it is a value for each node (for country drafting or army placing), a value for each edge (for attacking or fortifying) or another scalar or vector used for any other decision (value estimation for example).

Note that the presented design is flexible and can be adapted either at the graph convolutional layers that build the hidden board representation, the final layers that take the output of the convolutions and turn it into distributions over the nodes or edges, or even at the creation of the node tensors where node embeddings can be considered. The final model has 5 heads: one policy head for each decision (pick country, place armies, attack, fortify) and one value head to estimate the value of all players as a vector with 6 components. The network is designed so that the player with index 0 is the one to play, and every time the network is used, the boards are transformed accordingly by permuting the players indices in the board. We estimate the value of all players like in [6].

4.2 Loss Function and Further Parameters

The loss function used was the one detailed in [1]. It is composed of two terms related to the policy and value estimations. The target policy is the distribution induced by the visit counts of the MCTS at the root node. The target value is ideally a vector with 1 for the winning player and 0 for the rest, but for unfinished games we created an evaluation function that takes into account the number of armies a player would receive at the start of his turn. If P represents the neural network policy, z the target value vector and $\hat{z}$ the value estimation given by the network, the loss functions is given by

$$\mathcal{L} = -\sum_a \frac{N(s,a)}{N(s)} \cdot \ln P(a|s) - \sum_{i=1}^{6} z_i \ln [\hat{z}_i] + (1 - z_i) \ln [1 - \hat{z}_i]$$

where $N(s,a)$ is the visit count for action a from state s and $N(s)$ is the total visit count for state s.

Regarding the evaluation function used for unfinished games we normalized the values and made sure that the value of an unfinished board was always less than 0.9, keeping it far from 1 so that it was easier to differentiate between a good unfinished game and a winning one. This was done because in previous work, players often had problem finishing up games, even when they had completely winning positions. For the tree policy, we used the PUCT bandit formula from *AlphaGo*:

$$U(s,a) = Q(s,a) + c_b \cdot P(a|s) \frac{\sqrt{N(s)}}{N(s,a) + 1}$$

where $Q(s,a)$ is the mean action value estimated so far and c_b is a constant controlling the exploration.

For the optimizer, we chose the Adam optimizer [12] with a learning rate of 0.001 that decayed exponentially by a factor of 0.1 every 50 epochs. Regarding the *Expert Iteration* algorithm, we used 4000 simulations per expert labeling step. To speed up self play we chose moves by sampling the action proportional to the network policy without performing a search.

5   Experiments

In terms of gameplay, we made the following simplifications to the game:

1. Cards are only traded automatically for all players.
2. Attacks are always *till dead*, meaning that if an attack is chosen, it will only finish when either the attacker has only one army left or when the defender is eliminated and the attacker conquers the country.
3. The choice of how many armies to move to a conquered country was also eliminated. By default only one army is left in the attacking country, and the rest move to the newly conquered one. Fortifications work in a similar way, were all the armies that can be moved will be moved.
4. When placing armies, only one country is chosen and all armies are placed on it.

This simplifications were made to reduce the game complexity and test our first prototype on more basic game strategy. In addition to these gameplay changes, we also performed our tests on a simplified map.

Experiments were run on a Linux system with an AMD EPYC-Rome Processor (100 cores, 2 GHz each) and 120 GB of main memory. No GPU was used.

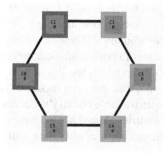

Fig. 2. Hex map. Country background colors are related to continents. Country names are $C0$ and $C1$ in red, $C2$ and $C3$ in yellow, $C4$ and $C5$ in blue (Color figure online)

5.1 Hexagonal Map

The hexagonal map shown in Fig. 2 has 6 countries grouped into three continents: *Red*, *Yellow* and *Blue*. The bonuses for each continent were designed so that their value is clearly different: *Red* has bonus 1, *Yellow* has bonus 4 and *Blue* has bonus 9. With this said, picking countries becomes very straightforward as the player must try to get *Blue* or at least stop the opponents from getting it. In a simple two player game, the optimal strategy for both players would end up with each player controlling one country of each continent so that the starting player does not have any bonus at his first turn. Moreover, if given the chance, a player should try to secure the *Blue* continent.

We started the evaluation of the models by looking at the evolution of the probabilities they assigned to each country when having to choose first. Results in Fig. 3 show that the model is indeed learning that the best choices are situated in the *Blue* continent, followed by the *Yellow* one. It is interesting to see that the model does not converge to a single choice. This might be useful to provide different lines of play if all options have indeed a similar value.

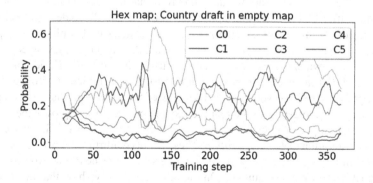

Fig. 3. Evolution of the policy for choosing a country with an empty board.

Next we wanted to know what happened if instead of choosing a country on an empty map, the model had to choose second and the best options were not available. Figure 4 shows that if the *Blue* country $C4$ was not available, the model is able to realize that the best option is either the other *Blue* country $C5$ to stop the enemy from getting the whole continent, or to start taking the *Yellow* continent. Again we see that the model changes its opinion through time between these two options. On the other hand, the probability of choosing the unavailable country is decreasing meaning the model is learning that the unavailable countries should not be picked. The model is also able to place armies in the most important countries after the country draft.

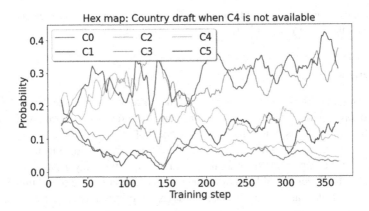

Fig. 4. Evolution of the policy for choosing a country after the best choice is not available

We tested the model against a random player by playing 100 games against it. We did this for each of the models after one iteration of *Expert Iteration* to see if there was an improvement during training. Our model only uses the neural net policy to weight the possible moves and sample one of them without performing any search. Moreover, having played against the model we had perceived that it struggled in the second part of the game, after country drafting and initial fortification. To understand better the performance of the model in the two parts of the game, we also played 100 games against random but using the player only in the initial part, and then playing completely at random. These two experiments are identified using the suffixes *full* and *init only*.

Figure 5 shows the evolution of the win rate for both experiments. To smooth the curves we present a rolling average with a window of 5 observations that better captures the tendency. It is interesting to see that the model considerably improves the winning rate of the random player after the initial phase (curve for *init only*). This is true even after only one iteration of *Expert iteration*. These results confirm what was found in [10] about UCT improving the country drafting, and also that our neural network is able to learn this UCT policy really quickly. On the other hand, they show that our player is worse than random in the second part of the game, and that even after 100 iterations of training it is not able to get the same win rate obtained by the player that uses the neural net policy in the initial phase and then random.

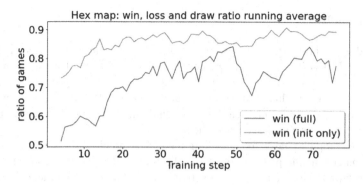

Fig. 5.

After having played against the trained model we can say that its behavior does not seem random or erratic. We think the player has trouble in some specific situations like finishing the game. Sometimes it will just stop attacking even when having an upper hand, and other times it will let the opponent recover and even become stronger by allowing it to keep continents that could (and should) be easily conquered. Some other times it will start accumulating armies without attacking even when it is obvious it should attack. All these behaviors seem to tell us that the attack phase needs further learning and that it is this phase that keeps the win rate of the player stuck below the random player. We would like to recall that no search was done, so the performance of the player might improve considerably if more resources are used and a search is performed using the policy instead of directly sampling a move using the policy.

6 Future Work

We think there are many possible improvements that could stem from this initial prototype. First, the game engine could allow faster or even parallel simulations to accelerate self-play and specially the expert labeling phase that involves the MCTS.

On the other hand there are many possible customizations for the neural network design to test. There are multiple options just when considering the graph convolutions to use (See for example [30, 31] for GNN and GCN reviews).

The training pipeline has also numerous parameters to study including the use of the value network in the MCTS [1, 25] or the way past experience is sampled [26]. In the paper presenting Katago [29], authors present interesting improvements to the *AlphaZero* pipeline that should also be beneficial for Risk like *Forced Playouts* and *Policy Target Pruning*.

Another natural step is to generalize the player to the full version of the game without simplifications. Some of the game phases would need to be reinterpreted, for example how to define a move in the army placing phase where N armies must be placed. In our opinion, the attack phase is the one that should be dealt with initially to really improve the player's level.

7   Conclusion

We created a first prototype of a Risk player using graph neural networks and MCTS following the *Expert Iteration* and *AlphaZero* frameworks. On small synthetic maps this model was able to learn optimal strategies for the country drafting phase, confirming that UCT can indeed produce a good country drafting policy [10] that can be learned by neural networks. We conjecture that this game phase might be somehow easier to learn because chance nodes appear deeper in the tree.

On more complex phases like attacking there were no clear signs of learning. The stochastic nature of this phase might be at the root of the problem and either more simulations have to be used to get better action value estimates or other techniques have to be thought of to deal specifically with chance nodes directly at the root.

Acknowledgment. This work was supported in part by the French government under management of Agence Nationale de la Recherche as part of the "Investissements d'avenir" program, reference ANR19-P3IA-0001 (PRAIRIE 3IA Institute).

References

1. Anthony, T., Tian, Z., Barber, D.: Thinking fast and slow with deep learning and tree search. arXiv preprint arXiv:1705.08439 (2017)
2. Anthony, T.W.: Expert iteration. Ph.D. thesis, UCL (University College London) (2021)
3. Auer, P., Cesa-Bianchi, N., Fischer, P.: Finite-time analysis of the multiarmed bandit problem. Mach. Learn. **47**(2), 235–256 (2002)
4. Blomqvist, E.: Playing the game of risk with an alphazero agent (2020)
5. Browne, C.B., et al.: A survey of monte carlo tree search methods. IEEE Trans. Comput. Intell. AI Games **4**(1), 1–43 (2012)
6. Carr, J.: Using graph convolutional networks and td (λ) to play the game of risk. arXiv preprint arXiv:2009.06355 (2020)
7. Cazenave, T.: Residual networks for computer go. IEEE Trans. Games **10**(1), 107–110 (2018)
8. Cazenave, T., et al.: Polygames: Improved zero learning. ICGA J. **42**(4), 244–256 (2020)
9. Coulom, R.: Efficient selectivity and backup operators Monte Carlo in tree search. In: van den Herik, H.J., Ciancarini, P., Donkers, H.H.L.M.J. (eds.) CG 2006. LNCS, vol. 4630, pp. 72–83. Springer, Heidelberg (2007). https://doi.org/10.1007/978-3-540-75538-8_7
10. Gibson, R., Desai, N., Zhao, R.: An automated technique for drafting territories in the board game risk. In: Proceedings of the AAAI Conference on Artificial Intelligence and Interactive Digital Entertainment. vol. 5 (2010)
11. Johansson, S.J., Olsson, F.: Using multi-agent system technologies in risk bots. In: Proceedings of the Second Artificial Intelligence and Interactive Digital Entertainment Conference (AIIDE), Marina del Rey (2006)
12. Kingma, D.P., Ba, J.: Adam: a method for stochastic optimization. arXiv preprint arXiv:1412.6980 (2014)
13. Kipf, T.N., Welling, M.: Semi-supervised classification with graph convolutional networks. arXiv preprint arXiv:1609.02907 (2016)
14. Kocsis, L., Szepesvári, C.: Bandit based Monte-Carlo planning. In: Fürnkranz, J., Scheffer, T., Spiliopoulou, M. (eds.) ECML 2006. LNCS (LNAI), vol. 4212, pp. 282–293. Springer, Heidelberg (2006). https://doi.org/10.1007/11871842_29

15. Li, G., Muller, M., Thabet, A., Ghanem, B.: Deepgcns: Can GCNS GO as deep as CNNS? In: Proceedings of the IEEE/CVF International Conference on Computer Vision, pp. 9267–9276 (2019)
16. Li, G., Xiong, C., Thabet, A., Ghanem, B.: DeeperGCN: all you need to train deeper GCNs. arXiv preprint arXiv:2006.07739 (2020)
17. Lütolf, M.: A Learning AI for the game Risk using the TD (λ)-Algorithm. Ph.D. thesis, BS Thesis, University of Basel (2013)
18. Melkó, E., Nagy, B.: Optimal strategy in games with chance nodes. Acta Cybernet. **18**(2), 171–192 (2007)
19. Nijssen, J., Winands, M.H.: Search policies in multi-player games1. J. Int. Comput. Games Assoc. **36**(1), 3–21 (2013)
20. Olsson, F.: A multi-agent system for playing the board game risk (2005)
21. Rosin, C.D.: Multi-armed bandits with episode context. Ann. Math. Artif. Intell. **61**(3), 203–230 (2011)
22. Scarselli, F., Gori, M., Tsoi, A.C., Hagenbuchner, M., Monfardini, G.: The graph neural network model. IEEE Trans. Neural Netw. **20**(1), 61–80 (2008)
23. Silver, D., et al.: Mastering the game of go with deep neural networks and tree search. Nature **529**(7587), 484–489 (2016)
24. Silver, D., et al.: A general reinforcement learning algorithm that masters chess, shogi, and go through self-play. Science **362**(6419), 1140–1144 (2018)
25. Silver, D., et al.: Mastering the game of go without human knowledge. Nature **550**(7676), 354–359 (2017)
26. Soemers, D.J., Piette, E., Stephenson, M., Browne, C.: Manipulating the distributions of experience used for self-play learning in expert iteration. In: 2020 IEEE Conference on Games (CoG), pp. 245–252. IEEE (2020)
27. Sturtevant, N.: A comparison of algorithms for multi-player games. In: Schaeffer, J., Müller, M., Björnsson, Y. (eds.) CG 2002. LNCS, vol. 2883, pp. 108–122. Springer, Heidelberg (2003). https://doi.org/10.1007/978-3-540-40031-8_8
28. Wolf, M.: An intelligent artificial player for the game of risk. Unpublished doctoral dissertation). TU Darmstadt, Knowledge Engineering Group, Darmstadt Germany (2005). http://www.ke.tu-darmstadt.de/bibtex/topics/single/32/type
29. Wu, D.J.: Accelerating self-play learning in go. arXiv preprint arXiv:1902.10565 (2019)
30. Wu, Z., Pan, S., Chen, F., Long, G., Zhang, C., Philip, S.Y.: A comprehensive survey on graph neural networks. IEEE Trans. Neural Netw. Learn. Syst. **32**(1), 4–24 (2020)
31. Zhang, S., Tong, H., Xu, J., Maciejewski, R.: Graph convolutional networks: a comprehensive review. Comput. Soc. Netw. **6**(1), 1–23 (2019). https://doi.org/10.1186/s40649-019-0069-y

Search in Games

Sequential Halving Using Scores

Nicolas Fabiano[1] and Tristan Cazenave[2(✉)]

[1] DI ENS, ENS Paris, Paris, France
nicolabiano22@yahoo.fr
[2] LAMSADE, Université Paris-Dauphine, CNRS, PSL, Paris, France
Tristan.Cazenave@dauphine.psl.eu

Abstract. We study the multi-armed bandit problem, where the aim is to minimize the simple regret with a fixed budget. The Sequential Halving algorithm is known to tackle it efficiently. We present a more elaborate version of this algorithm to integrate some exterior knowledge or "scores", that can for instance be provided by a neural network or a heuristic such as all-moves-as-first (AMAF) in the context of a Monte-Carlo Tree Search. We provide both theoretical justifications and experiments.

1 Introduction

Since it was introduced in [6, 11], the Monte Carlo Tree Seach (MCTS) algorithm has known a great success in AI, especially in turn-based games like Go, and some of its refinements are state of the art for most games.

The general idea of this algorithm is the following: from the root configuration, it picks a move, and generates a random playout from it. If the player to move wins, this means that the move was probably good, and if they lose, it was probably bad. Then the algorithm continues by picking more moves, deeper and deeper in the game tree, respecting a fixed amount of playout (or time) budget.

One of the key elements for MCTS to be efficient is the choice of what moves to investigate, with the usual search for the optimal exploration-exploitation trade-off. To perform this, one typically uses the Upper Confidence Bound (UCB) bandit algorithm, which has good properties in terms of cumulative regret. This means that, for every investigated configuration, the moves tested were mostly good ones.

However, in the context of games, the success of simulations does not matter in itself. The only goal is that the final output of the algorithm is as good a move as possible. This means that, instead of cumulative regret, a more relevant quantification is the expected simple regret (see Fig. 1 for a precise definition).

In [10], a new bandit algorithm named Sequential Halving (SH) was introduced. It is proven to have a small expected simple regret 0-1 (see Fig. 1) and is also shown to have a small expected simple regret through numerical experimentation. It has successfully been used as an alternative to UCB in MCTS, in particular as a replacement in the root node with UCB used in the rest of the tree [12], in Partially Observable Games [13] or even in the whole tree with SHOT [3].

However, for most games, the unmodified UCB is not state of the art. For many games such as Go, moves typically commute, so the RAVE algorithm, which uses the

© Springer Nature Switzerland AG 2022
C. Browne et al. (Eds.): ACG 2021, LNCS 13262, pp. 41–52, 2022.
https://doi.org/10.1007/978-3-031-11488-5_4

all-moves-as-first (AMAF) heuristic [1], was introduced [9]. For some games, once again including Go [14], Neural Networks (NN) can provide an algorithm with reliable priors, which are incorporated in the PUCT algorithm [14].

The aim of this paper will be to incorporate exterior knowledge like AMAF or NN to the SH algorithm, and to compare the result both to the simple SH and to the state of the art MCTS algorithms RAVE and PUCT.

The first part will discuss the SH algorithm in general, and report experiments in a theoretical setup. The second part will present a theoretical foundation for a new algorithm named SHUSS, Sequential Halving USing Scores. It will also discuss some variations around it, and report experiments on games.

With p_i the mean of arm i, i^* the optimal arm and $\hat{i}$ the chosen one,

Cumulative regret:	Simple regret:	Simple regret 0-1:
$\mathcal{R}_{\text{cum}} = \sum_{r \text{ round}}(p_{i^*} - p_r)$	$\mathcal{R} = p_{i^*} - p_{\hat{i}}$	$\mathcal{R}_{0-1} = 1$ if $i^* \neq \hat{i}$ else 0

Fig. 1. The various notions of regret

2   The Sequential Halving Algorithm

The SH algorithm is round-based. For every round, each arm is sampled the same amount of times, and only a set fraction of the best arms are kept. This step is repeated until there is only one arm left.

The theoretical bounds presented in [10] suggest that the same total budget should be spent for each round, and that the fraction removed should be constant for every step (denoted $1 - \lambda$). For a precise description, see Algorithm 1.

This version of the algorithm differs from the original one in two ways. First, by the introduction of the parameter λ, which allows for values other than the original 1/2. Second, the computation of the budget per round is slightly improved, to ensure that less budget is left unspent in case of multiple issues with rounding.

Note 1. Contrary to other bandit algorithms like UCB, SH assigns a lot of the budget at once to each arm, which has practical advantages like simpler parallelisation and less back-and-forth in the search tree. This is especially true when λ is small (few rounds).

2.1   Restart vs Stockpile

In [10], for the theoretical computations to be rigorous, one has to assume that rounds are independent, which means that statistics are discarded from one round to the other.

However, in order to gather more accurate statistics, it may be worth to *stockpile* the statistics from the previous round, instead of *restarting* them for every round. In terms of budget, this adds a factor of almost $1/(1 - \lambda)$.

Note 2. Getting the factor of almost $1/(1 - \lambda)$ from the first rounds implies redistributing the weight to give more of it at the beginning, but less at the end. Doing this will be referred to as *uniforming*.

Algorithm 1. Sequential Halving

Parameter: cutting ratio λ
Input: total budget T, set of arms S
$S_0 \leftarrow S, T_0 \leftarrow T$
$R \leftarrow$ number of rounds before $|S_R| = 1$
for $r = 0$ **to** $R - 1$ **do**
  $t_r \leftarrow \lfloor \frac{T_r}{|S_r| \cdot (R-r)} \rfloor$
  $T_{r+1} \leftarrow T_r - t_r |S_r|$
  sample t_r times each arm in S_r
  $S_{r+1} \leftarrow S_r$ deprived of the fraction $1 - \lambda$ of the worst arms
end for
Output: arm in S_R

In theory, this may cause the following issue: if, for one round, a rather bad arm is sampled disproportionately, these statistics will be stockpiled for the next round, which will cause it to be kept even further; whereas restarting would decrease the probability for that bad arm to be chosen, as it would have to be wrongly selected twice. This issue is particularly important when λ is close to 1, as the stockpiled statistics contribute significantly to the overall ones in that case.

A compromise can be found between the two pure approaches, as one can keep the statistics from the previous round and give it a decaying factor d between 0 (pure stockpile) and 1 (pure restart).

The experiments of the next section clearly show that stockpiling is always better, even more so than choosing $0 < d \ll 1$.

Note 3. We successfully replicated the SH part of the experiments of [10], and it would appear that they were done using stockpiling, as restarting gives significantly worse results.

2.2 Experiments

Even if we could be more general, we focus on the case where the only possible outcomes are 0 (loss) and 1 (win). Thus, every arms is described by its *value*, which is both the probability to win and the expected value.

The performance of bandit algorithms highly depend on the distribution of the arms' values. We consider 4 distributions of values for the n arms.

In setting (1), the optimal arm has a value of 0.5 and the others have a value of 0.4.

In setting (A), the values form an arithmetic sequence from 0.5 to 0.25.

In setting (S), the optimal arm has a value of 0.5, the worst has a value of 0.25, and the others have values such that i/δ_i^2 is constant, with δ_i the difference in value with the optimal arm. This setting is suggested by the fact that the theoretical bounds of [10] rely on these values, and thus the theoretical guaranty is the strongest.

In setting (N), the values are distributed according to the sigmoid of a normal distribution with parameters 0.5 and $\sigma^2 = 0.01$. This setting induces richer behaviours, and we believe it to be a more realistic model of the actual distributions in games.

The results are compared to UCB, the standard MCTS bandit. It consists of, for each step from 1 to the budget, picking the arm that maximises the empirical value, added to a term to force exploration, of the form

$$c\sqrt{\frac{\log(\text{playouts})}{\text{playouts}_z}} \tag{1}$$

We tested various values for λ and d for SH, and compared it to various values for the exploration constant c in UCB. We also tested the uniforming variant discussed in Note 2. The results are shown in Fig. 2.

Rounding the number of arms left is handled as follows: always round up, except when this would cause the amount of arms to remain constant, in which case round down.

Each result is averaged over 10000 tests. To reduce the covariance from one setting to another, the bandits are seeded using *numpy.random.binomial*. For the same index of experiment e and the same arm i, if the value of arm i is the same in two settings, then on the same round r their results are drawn out of the same sequence of win/loss (the number of successes is monotonous in terms of budget).

As announced, in every setting, the best results are obtained for $d = 0$, showing that in practice, stockpiling is more efficient than restarting.

The optimal λ depends on the setting. The experiments globally suggest that, for the interesting case $d = 0$, $\lambda \approx 0.7$ is often the best value, but the difference is small and the algorithm performs well on a wide range of λ that includes the classical value $\lambda = 0.5$. That problem is actually very complex, and some less rigorous experiments suggest that it is better not to decrease geometrically but rather to start with large decreasing factors and to end with smaller ones (eg $20 \rightarrow 8 \rightarrow 4 \rightarrow 3 \rightarrow 2 \rightarrow 1$ rather than $20 \rightarrow 10 \rightarrow 5 \rightarrow 3 \rightarrow 2 \rightarrow 1$).

The effect of uniforming is mixed, which suggests that there is room for practical improvement concerning the way the budget is distributed among the rounds.

Surprisingly, the results are globally worse than UCB for $n = 20$, especially in the setting (S), for which the SH algorithm is theoretically designed. Nonetheless, UCB relies more heavily on fine-tuning of its parameter c, with no universally excellent value, and for $n = 80$ SH is globally better.

3   Scores

The aim of this part will be to develop a variant of the SH algorithm that takes advantage of some exterior knowledge, like a NN or AMAF statistics. We will consider the general case where we have access to what we will call a *score*, which is a numerical evaluation of every move, independent from the bandit evaluation.

The bandits are still assumed to give either 0 or 1, giving an empirical mean $p_r^{(i)} \in [0, 1]$ for arm i on round r, but the scores do not necessarily belong to $[0, 1]$.

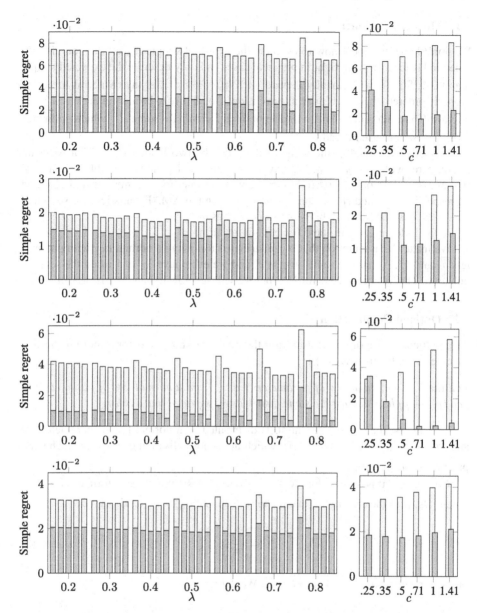

Fig. 2. Simple regret obtained with SH in various settings. In every setting, the budget is taken equal to $T = 2048$. From top to bottom, we report settings (1), (A), (S) and (N). For each setting, the left plot corresponds to SH, and the right one corresponds to UCB. For SH, for each λ, the bars correspond (from left to right) to $d = 1$, $d = 0.5$, $d = 0.1$, $d = 0$, and $d = 0$ with uniforming. The darker bars correspond to $n = 20$, and the lighter ones to $n = 80$.

3.1 Theoretical Model

We don't know precisely how to estimate the expected simple regret: the bounds provided in [10] are far from tight in practical cases and only describe the expected simple regret 0-1. Still, it will globally depend on $P(p_r^{(i)} < p_r^{(j)})$: if any two arms are often properly ordered, then the best arms have a low probability to be among the worst $1 - \lambda$ fraction. Thus, our aim will be to find an optimal formula for some $q_r^{(i)}$ which optimizes $P(q_r^{(i)} < q_r^{(j)})$ to use instead.

Formally, let x and y (the value of the arms) be two hidden values that we want to compare, with $x - y = \delta$. We have access to 4 independent variables. X and Y (the number of 1s obtained) are binomials with a same first parameter t and centered on respectively tx and ty. $\tilde{X}$ and $\tilde{Y}$ (the scores, eg the AMAF statistics) are such that $\tilde{X} - \tilde{Y} = \tilde{\delta}$ is hopefully globally the same sign as δ.

In the following, z can stand for x, y, or any arm.

We make the assumption that $\tilde{\delta}$ is distributed following a normal law with parameters $\tilde{\delta}_0$ and $\tilde{\sigma}_0^2$. $\tilde{\delta}_0$ has the same sign as δ, and we even have $\tilde{\delta}_0 = \delta$ when the score is unbiased. This is not the case for NN, but we will see how to handle this in Sect. 3.5.

3.2 Optimal Combination

As a particular case of the central limit theorem, we know that (for a more quantified statement, see for instance [7]):

Lemma 1. *A binomial law of parameters t and p and a normal law of parameters tp and $tp(1 - p)$ have almost the same distribution, provided that t is large.*

This means that $X - Y$ is (almost) distributed as a normal law of parameters $t\delta$ and $t\sigma^2 = t(x(1 - x) + y(1 - y))$, which up to normalisation can be seen as having parameters $\tilde{\delta}_0$ and $\frac{\tilde{\delta}_0^2 \sigma^2}{\delta^2 t}$.

Conversely, this shows that $\tilde{X} - \tilde{Y}$ gives (almost) the same information as two binomials, with the crucial first parameter $\tilde{t}$ such that $\tilde{\sigma}_0^2 = \frac{\tilde{\delta}_0^2 \sigma^2}{\delta^2 \tilde{t}}$. This gives

$$\tilde{t} = \frac{\tilde{\delta}_0^2 \sigma^2}{\delta^2 \tilde{\sigma}_0^2} \tag{2}$$

but with an intensity $\frac{\tilde{\delta}_0}{\delta}$ that is too large. We define

$$\tilde{t}' = \frac{\tilde{\delta}_0 \sigma^2}{\delta \tilde{\sigma}_0^2} \tag{3}$$

We showed that the problem is (almost) equivalent to maximizing the probability of choosing the best arm among two, knowing that one has succeeded $X + \tilde{t}'\tilde{X}$ times out of $t + \tilde{t}$ trials, and the second $Y + \tilde{t}'\tilde{Y}$ times.

Thus, it is optimal to use (for $\frac{\tilde{\delta}_0^2 \sigma^2}{\delta^2 \tilde{\sigma}_0^2}$ reasonably large)

$$q_z = Z + \tilde{t}'\tilde{Z} \tag{4}$$

Similar reasoning gives the same result for t reasonably large.

One could be tempted to use the $\tilde{Z}$ to approximate σ. However, given that the final goal is to sort all the arms on a single scale, $\tilde{t}'$ has to be the same for every pair of arms.

The simplest solution is to choose a hyperparameter $\tilde{t}'$ that corresponds to an overall reasonable guess. We will see how to improve that choice in some particular cases.

The resulting algorithm is presented as Algorithm 2. In it, t_r^+ corresponds to the total budget used in $p_r^{(i)}$: $t_r^+ = t_r$ with restart and $t_r^+ = t_0 + \cdots + t_r$ with stockpile.

Algorithm 2. Sequential Halving USing Scores (SHUSS)

Parameter: cutting ratio λ, $\tilde{t}'$
Input: total budget T, set of arms S, online scores $\tilde{X}_r^{(i)}$
$S_0 \leftarrow S, T_0 \leftarrow T$
$R \leftarrow$ number of rounds before $|S_R| = 1$
for $r = 0$ **to** $R - 1$ **do**
$\quad t_r \leftarrow \lfloor \frac{T_r}{|S_r| \cdot (R-r)} \rfloor$
$\quad T_{r+1} \leftarrow T_r - t_r |S_r|$
$\quad$ sample t_r times each arm in S_r, giving an empirical mean $p_r^{(i)}$ to arm i out of t_r^+ trials
$\quad q_r^{(i)} = p_r^{(i)} + \frac{\tilde{t}'}{t_r^+} \tilde{X}_r^{(i)}$
$\quad S_{r+1} \leftarrow S_r$ deprived of the fraction $1 - \lambda$ of the worst arms in terms of $q_r^{(i)}$
end for
Output: arm in S_R

3.3 Selection Bias

One issue that may occur is that, after any given round, the arms that remain have their $\tilde{Z}$ biased by the fact that they were among the best. Thus, even if during the first round they are indeed normal laws, it is unclear how they look like after a few rounds.

However, this issue is very similar to the issue of stockpiling, as all arms tend to have better stats than they theoretically should. The fact that stockpiling is so powerful suggest that this issue is not too important, so we will neglect it.

3.4 Case of AMAF: a Better Formula for $\tilde{t}'$

This subsection discusses the special case where the scores are given by AMAF statistics. It should be seen as a small toolbox consisting of a few ideas that can be used to do better than taking $\tilde{t}'$ as a constant, based on a case study.

The AMAF (all-moves-as-first) score [1] consists in evaluating a move m for a player p in a game state s, considering the win/loss ratio of every game where p plays m, not only in s itself but in any of its descendants in the game tree (or even its cousins, in some variants of AMAF like GRAVE [2]).

First of all, this score is not independent from the value of the bandits. In the first rounds of the algorithm, there are many bandits, so the AMAF scores are almost independent from each of them, which makes it a mostly unimportant issue.

In the last rounds, however, it is highly correlated with the stats of some, if not all, bandits. In some games, one could imagine that some properties of the moves generate important biases, for instance if the move m can only appear after few of the remaining moves considered. We will see a general way to address this problem, but this could be more tricky for some particular games and we recommend caution.

The most interesting aspect about AMAF in this context is that the score becomes more and more accurate as evaluations are performed. Thus, taking $\tilde{t}'$ as a constant throughout the algorithm is not adequate. Instead, one can model the distribution of $\tilde{\delta}$ as based on the following:

- the fact that AMAF is a heuristic causes an error distributed as a normal law of variance σ_{heu}^2, centered somewhere between δ and the local average value;
- the fact that the AMAF stats are only gathered on a finite number s_r of moves on round r causes an error distributed as a binomial law, which is almost (see Lemma 1) and after normalization a centered normal law of variance $\sigma_{\text{stat}}^2/s_r$.

Provided that σ_{heu}^2 is small (i.e. the heuristic makes sense in the application context), and the values of the arms are not too extreme, σ and σ_{stat} are almost equal.

Equation 3 applied with this variance gives

$$\tilde{t}'_r = \frac{\tilde{\delta}_0}{\delta} \frac{\sigma^2}{\sigma_{\text{heu}}^2 + \sigma_{\text{stat}}^2/s_r} \approx \frac{\tilde{\delta}_0}{\delta} \frac{1}{\sigma_{\text{heu}}^2/\sigma^2 + 1/s_r} \tag{5}$$

This time, there are 2 hyperparameters to choose values for.

$\frac{\tilde{\delta}_0}{\delta}$ describes how much AMAF flattens the stats, and can easily be measured experimentally. It may be relevant to make it depend on the number of arms left and on the variant of AMAF used.

$\sigma_{\text{heu}}/\sigma$ describes how accurate the heuristic is, compared to the accuracy provided by binomial stats. Giving this hyperparameter a relatively high value also ensures that, in the last rounds where s_r is large, the value of $\tilde{t}'_r$ stops increasing, which addresses the previously mentioned issue of correlation.

Note that this reasoning works only if, on each round, s_r is globally the same for every arm (or if, for every arm, $1/s_r \ll \sigma_{\text{heu}}^2/\sigma^2$), as we need a common value of $\tilde{t}'_r$.

3.5 Case of Prior Score: Pruning

In this subsection, we assume that the $\tilde{X}_i$ are known a priori (before any evaluation is performed). This can be applied to some extent in cases where some score is known a priori but refined during the algorithm, like GRAVE. We start with a general discussion, before dealing with the specific case of neural networks (NN) applied to MCTS.

Even before the algorithm begins, some arms have no chance of being chosen at the end, for instance if $\tilde{X}_i$ is smaller than the median (for $\lambda = 1/2$) minus $1/\tilde{t}_0'$.

In addition to these trivial pruning operations, it is often worth pruning more arms, as the budget saved will compensate for the risk taken.

As we saw in a previous section, the prior can be interpreted as though we had already spent some amount $\tilde{t}$ of budget on each arm before round 0, which we will consider to be a round labelled -1. The philosophy of SH (exploited in the performance proof in [10]) is that, when bandits are pruned up to number n_r with a budget t_r, the product $\pi_r := n_r \cdot t_r$ is equal to some π that does not depend on r. Thus, it is natural to prune up to arm n_{-1}, where n_{-1} is chosen so that $\pi_{-1} = \pi$.

For a precise computation, we neglect the rounding issues when dividing by λ. We also make the computations as if we were not stockpiling (note that using the score on the subsequent rounds can be seen as stockpiling when it is purely a prior). Then

$$\pi_{-1} = n_{-1} \cdot \tilde{t} \tag{6}$$

$$\pi = \pi_0 = \lambda n_{-1} \cdot \frac{T}{\log_{1/\lambda}(n_{-1}) \cdot n_{-1}} \tag{7}$$

$$n_{-1} \log_{1/\lambda}(n_{-1}) = \frac{\lambda T}{\tilde{t}} \tag{8}$$

For NN in MCTS, all the previous theoretical foundation has to be slightly adapted, given that bandits don't give 0 or 1 but the value of the leaf evaluated by the NN instead.

More importantly, the score is given by the policy of the root. It is meant to be monotonic with the value, but the way it uses a softmax layer makes the rest of our model about its distribution fail. Thus, the safest way to use it is for pruning, and then the remaining arms are explored using a basic SH that does not use the policy.

The previous formula for n_{-1} should hold for the same reasons. Now, the value of $\tilde{t}$ describes how much budget is needed for the exploration to be as good as the policy. As the budget is typically distributed among the rest of the tree by an algorithm like PUCT, designed to be good asymptotically but not for small values, $\tilde{t}$ is typically quite large. In addition, given that the policy is not stockpiled, it is better to overestimate the value of $\tilde{t}$ to make use of the policy as much as possible.

3.6 Experiments with AMAF

First, we test SHUSS using the score AMAF, to compare it with RAVE [8,9].

The latter uses the AMAF score as follows: the value of the arm, to which the exploration term is added, is taken equal to

$$(1 - \beta_z)Z + \beta_z \tilde{Z} \tag{9}$$

with t_z the number of playouts starting with z, s_z the number of playouts containing z and

$$\beta_z = \frac{s_z}{s_z + t_z + bias \times s_z \times t_z} \tag{10}$$

Table 1. Percentage of games won by Hybrid-SHUSS (using AMAF and RAVE) against RAVE, in various games. $T = 10000$; $bias = 10^{-7}$; $\lambda = 1/2$; 500 matches.

Game \ $\tilde{t}'$	0	128	256	512	1024	2048	4096	8192	16384	∞
Atarigo 7 × 7	44.2	47.2	49.6	**50.2**	50.0	49.6	45.2	47.8	46.4	45.2
Atarigo 9 × 9	35.6	41.4	40.0	38.2	41.0	41.2	**43.4**	41.4	36.4	40.2
Ataxx 8 × 8	30.2	33.6	35.2	34.2	42.0	46.2	55.0	62.4	62.0	**71.8**
Breakthrough 8 × 8	54.0	**57.8**	56.8	56.0	56.6	55.2	53.8	51.0	55.0	52.4
Domineering 8 × 8	41.4	47.8	44.8	**49.0**	46.2	47.2	46.2	45.6	43.0	42.4
Go 7 × 7	45.2	49.2	46.2	53.8	**58.6**	50.2	42.6	33.2	31.0	15.8
Go 9 × 9	43.4	53.2	**58.2**	52.2	50.8	43.8	35.6	26.4	19.0	12.2
Hex 11 × 11	15.8	43.0	43.4	**51.4**	48.4	50.2	46.4	46.6	43.4	42.6
Knightthrough 8 × 8	61.0	61.6	**65.0**	63.8	62.2	60.2	54.2	54.4	56.2	52.8
NoAtaxx 8 × 8	**91.0**	87.4	76.8	72.0	62.8	55.2	53.8	44.6	45.8	43.2
NoBreakthrough 8 × 8	37.8	40.8	44.0	46.2	**51.4**	44.2	46.4	44.0	50.0	46.6
NoDomineering 8 × 8	40.4	45.6	49.4	46.0	48.4	**50.0**	47.6	47.4	45.0	47.6
NoGo 7 × 7	38.8	40.8	45.6	44.0	50.8	47.6	50.8	49.4	47.6	**51.8**
NoGo 9 × 9	30.0	37.8	38.8	40.0	41.0	42.0	42.8	45.0	**45.8**	37.4
NoHex 11 × 11	46.4	48.0	48.6	49.0	**49.2**	48.6	48.6	49.2	48.8	49.2
NoKnightthrough 8 × 8	29.0	36.8	38.8	39.6	47.8	46.2	46.0	45.2	**48.2**	47.6
Average	42.76	48.25	48.83	49.10	**50.45**	48.60	47.40	45.85	45.23	43.68

Table 2. Percentage of games won by Hybrid-SHUSS (using a NN and PUCT) against PUCT, for the game of Go. $c = 0.2$; $\lambda = 1/2$; 400 matches.

$T \setminus n_{-1}$	3	4	5	6	7	8	9
32		31.00	**46.00**	43.50	26.50	20.50	
64		57.75	60.00	57.00	**71.50**	38.75	
128		39.75	46.50	**54.50**	39.75	41.25	
256				55.25	**71.50**	60.75	
512				25.50	**60.00**	47.25	
1024					60.75	**67.75**	55.50

[12] demonstrates how to combine the SH algorithm with UCT in the Hybrid-MCTS algorithm: SH is used only at the root, and the rest of the tree expansion uses UCB. We followed this idea, by combining SHUSS at the root with RAVE for the rest of the tree, in an algorithm naturally named Hybrid-SHUSS.

Table 1 reports the results of 500 matches (250 as White and 250 as Black) between Hybrid-SHUSS and RAVE, for many classical games. Both algorithms use a budget (number of playouts) per move equal to 10000. RAVE uses the classical parameter $bias = 10^{-7}$, both in the inner parts of Hybrid-SHUSS and its opponent. SHUSS uses the classical parameter $\lambda = 1/2$.

Different values of $\tilde{t}'$ are experimented with (to keep things simple, $\tilde{t}'$ is a constant). The extreme case $\tilde{t}' = 0$ is the usual SH algorithm without AMAF (it is only used to break ties), and $\tilde{t}' = \infty$ is relying purely on AMAF, with the same weight regardless of whether or not the move is first.

In most games, SHUSS performs better than both pure SH and pure AMAF.

The optimal value of $\tilde{t}'$ depends on the game, but using 1024 gives a reasonably good performance for every game with this budget.

3.7 Experiments with a Neural Network

We then test SHUSS with a prior given by a NN in the game of Go.

The state of the art NNs in the game of Go use two heads, one for the policy and one for the value. The MCTS algorithm used in current computer Go programs since AlphaGo is PUCT. It uses the NN score as follows : the exploration term is replaced by

$$c \times \tilde{Z} \times \frac{\sqrt{t}}{1 + t_z} \tag{11}$$

with $\tilde{Z}$ the policy, t the budget already used for this node and t_z the budget already used for this node on move z.

We use as a NN a simple MobileNet of 16 blocks, a trunk of 64 and 384 planes in the bottleneck block [4,5]. MobileNets give better results than usual residual networks for the game of Go.

As explained in Sect. 3.5, in SHUSS, the policy is used to prune at the root, and the remaining algorithm is SH at the root and PUCT for the remaining of the tree, in a similar Hybrid fashion.

Experiments showed that PUCT performs best against Hybrid-SHUSS for $c = 0.2$. Table 2 report the results of 400 matches of Hybrid-SHUSS against PUCT for various budgets and n_{-1}, and we see that with the Hybrid-SHUSS outperforms PUCT for large enough budgets.

Concerning the relationship between T and the optimal n_{-1}, it seems to be logarithmic, while it was theoretically expected to be closer to being linear. Regardless, the size of the game of Go forced us to stick to rather small budgets, for which this part of the theory may not apply yet as it is asymptotic.

4   Conclusion

In the first section, we discussed the SH algorithm in general.

We discussed two ways of using the budget, restarting and stockpiling, with the latter being much better experimentally.

We also showed that a cutting parameter $\lambda \approx 0.7$ for SH is experimentally slightly better than the classical $\lambda = 0.5$, but that globally the algorithm is very robust for a wide range of λ. Nonetheless, it appears that some more flexible budget attribution or cuts may be better.

In the second section, we presented our new algorithm Sequential Halving USing Scores (SHUSS).

A theoretical model suggests a very simple way to combine the score with the bandit statistics, while still leaving plenty of room for improvement depending on the precise nature of the score.

Work still remains to be done to handle scores that are very asymmetrical among the arms in terms of quality.

We associated SHUSS with AMAF statistics and RAVE under the root play in a variety of different games against RAVE with the same parameters. The results are mixed, and depend on the game.

We also made SH, pruning using the policy with PUCT under the root node, play Go against PUCT with the same parameters. SH with pruning outperforms PUCT for well chosen numbers of moves kept, but it is quite sensitive to this value and it is unclear how to choose it in general.

Acknowledgment. This work was supported in part by the French government under the management of the Agence Nationale de la Recherche as part of the "Investissements d'avenir" program, reference ANR19-P3IA-0001 (PRAIRIE 3IA Institute).

References

1. Bouzy, B., Helmstetter, B.: Monte-Carlo Go developments. In: ACG, pp. 159–174 (2003)
2. Cazenave, T.: Generalized rapid action value estimation. In: IJCAI, pp. 754–760 (2015)
3. Cazenave, T.: Sequential halving applied to trees. IEEE Trans. Comput. Intell. Games **7**(1), 102–105 (2015)
4. Cazenave, T.: Improving model and search for computer Go. In: IEEE CoG (2021)
5. Cazenave, T.: Mobile networks for computer Go. IEEE Trans. Games **14**, 76–84 (2021)
6. Coulom, R.: Efficient selectivity and backup operators in Monte-Carlo tree search. In: Computers and Games, pp. 72–83 (2006)
7. Sunklodas, J.K.: On the normal approximation of a binomial random sum. Lithuanian Mathematical Journal **54**(3), 356–365 (2014). https://doi.org/10.1007/s10986-014-9248-6
8. Gelly, S., Silver, D.: Combining online and offline knowledge in UCT. In: ICML, pp. 273–280 (2007)
9. Gelly, S., Silver, D.: Monte-Carlo tree search and rapid action value estimation in computer Go. Artif. Intell. **175**(11), 1856–1875 (2011)
10. Karnin, Z.S., Koren, T., Somekh, O.: Almost optimal exploration in multi-armed bandits. In: ICML, pp. 1238–1246 (2013)
11. Kocsis, L., Szepesvári, C.: Bandit based Monte-Carlo planning. In: ECML, pp. 282–293 (2006)
12. Pepels, T., Cazenave, T., Winands, M.H.M., Lanctot, M.: Minimizing simple and cumulative regret in Monte-Carlo tree search. In: Cazenave, T., Winands, M.H.M., Björnsson, Y. (eds.) CGW 2014. CCIS, vol. 504, pp. 1–15. Springer, Cham (2014). https://doi.org/10.1007/978-3-319-14923-3_1
13. Pepels, T., Cazenave, T., Winands, M.H.M.: Sequential Halving for Partially Observable Games. In: Cazenave, T., et al. (eds.) CGW/GIGA -2015. CCIS, vol. 614, pp. 16–29. Springer, Cham (2016). https://doi.org/10.1007/978-3-319-39402-2_2
14. Silver, D., et al.: Mastering the game of Go with deep neural networks and tree search. Nature **529**(7587), 484–489 (2016)

Cosine Annealing, Mixnet and Swish Activation for Computer Go

Tristan Cazenave$^{(\boxtimes)}$, Julien Sentuc, and Mathurin Videau

LAMSADE, Université Paris-Dauphine, PSL, CNRS, Paris, France
Tristan.Cazenave@dauphine.psl.eu

Abstract. The architecture of neural networks in neural based computer game programs influences greatly the strength of the game playing programs. We present developments on the recently tested Mobile Network architecture that has good results for the game of Go. The three proposed improvements deal with the optimization process, the activation function and the convolution layers. These three modifications improve the accuracy of the policy and the error of the evaluation, as well as the playing strength of a computer Go program using the resulting networks.

1 Introduction

Important breakthroughs in Computer Go have been achieved in the past years. These advances were made possible by the advent of Convolutional Neural Network (CNN) and development of Monte-Carlo Tree Search. Because of their versatility, CNN architectures are constantly evolving. Thus, the purpose of this article is to use these recent changes to improve supervised learning in Computer Go. We also hope that these improvements will transfer to the Reinforcement Learning setup.

Classically, CNN for Go have more than one head. At least, these networks use a policy head, to prescribe moves, and a value head, to evaluate the board quality in terms of future incomes. This output configuration has been popularized by the groundbreaking AlphaZero [7]. In 2017, AlphaGoZero reached a superhuman level without initial knowledge except the game rules. Thereafter, DeepMind's algorithm AlphaZero achieved comparable results for Chess and Shogi. This achievement has been made possible by the use of deep reinforcement learning from self-play.

Closer to our work, various architecture has been evaluated for learning to play Go in a supervised way. Typically, the dataset used for the supervised learning is constituted of superhuman games produced by Deep Reinforcement Learning agent like AlphaZero or Katago. In our study, we used Katago to constitute our dataset.

KataGo [9] like Alpha Zero only learns from neural-net-guided Monte Carlo Tree Search self-play. KataGo improves learning compared to AlphaGo Zero. Mainly, it converges to superhuman level much faster than comparable methods

© Springer Nature Switzerland AG 2022
C. Browne et al. (Eds.): ACG 2021, LNCS 13262, pp. 53–60, 2022.
https://doi.org/10.1007/978-3-031-11488-5_5

such as Alpha Zero, ELF/OpenGo or Leela Zero. It uses different optimizations strategies like using a low number of playouts for most of the moves in a game to gather more data about the value in a shorter time, or using additional training targets to regularize the network. An innovation in the Katago program is to use GlobalAverage Pooling in some layers of the network in conjunction with residual layers.

Architectures of the Neural Network used in Deep Reinforcement Learning has been shown to have a great impact on the performances of the resulting playing engines. For example, the use of residual networks increased Alpha GO's ELO by 600. Residual Networks used in Alpha Zero were compared to Mobile Networks [4] with policy and value heads different from the Alpha Zero ones, for instance a fully convolutional policy head and a global average pooling value head. These mobile networks are more efficient in terms of computation and parameters than their classic CNN counterparts. Also, further improvement of mobile networks have been tested [3]. The main architecture change is the introduction of the Squeeze and Excitation block, adding channel attention to the network. Mobile Networks presented better results than Residual Networks, both for small and large networks on the Leela dataset composed of games played at a superhuman level [4]. They had a better accuracy and value error.

2   Improving Supervised Learning

Here, we present different improvements made to increase performance. They rely on three different aspects of the training: optimization, activation function and architecture.

2.1   Cosine Annealing

Better optimization schema can lead to better results. Indeed, by using a different optimization strategy, a neural net can end in a better optimum. In this paper, this is achieved by using Stochastic Gradient Descent with warms Restart (SGDR) [5]. In particular, the learning rate is restarted multiple times. This way, the objective landscape is explored further and the best solution of all restart is kept. Furthermore, using a peculiarly aggressive learning rate strategies like cosine annealing (Eq. 1) can lead to better convergence.

$$\eta_t = \eta_{\min}^i + \frac{1}{2}(\eta_{\max}^i - \eta_{\min}^i)(1 + \cos(\frac{T_{cur}}{T_i}\pi)) \tag{1}$$

with η_t the learning rate at time t, T_{cur} the number of step since the last restart and i the current number of cycles done. Thus, T_i indicates the number of steps allowed for the cycle i and $\eta_{\min}^i, \eta_{\max}^i$ the range of values the learning rate can take during the cycle i.

We compare this cosine annealing with what we call *division annealing*. Division annealing, simply divide the learning rate by a constant at predefined epochs.

2.2   MixNet

Traditional depthwise convolution suffers from the limitations of single kernel size. *Tan et al.* [8] proposed to replace the vanilla depth-wise convolution with *MixConv*. Their module takes advantage of bigger kernel size in the convolution. The idea is to mix up multiple kernels of different sizes in a single depthwise convolution operator in order to capture different types of patterns at different scales from input images. They achieved significant performance gain in image classification compared to mobilenet-v3 on both ImageNet classification and COCO object detection. Even better, they showed that mixing kernel size allows using bigger kernels.

2.3   Swish Activation

Non-linearity plays an important role in neural network. Without them, they lose their expressiveness power. It also has an important impact on the neural net training. In particular, the activation shape the derivatives of the network. These important properties motivated *Ramachandran et al.* [6] to search for good activation functions. From their research, they discovered the Swish activation function:

$$x \cdot \text{sigmoid}(x)$$

This activation, used as drop down replacement for ReLU, gives significant improvement on diverse tasks and networks (Fig. 1).

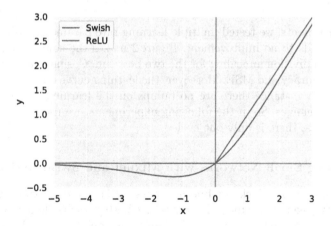

Fig. 1. Swish function plot

3   Experimental Results

In all experiments, instead of learning the final result of a game, the value head is labeled with the Q value coming from MCTS. We use a simple mixed convolution composed of half of 3×3 kernels and half of 5×5 kernels. The dataset is composed of self played games from Katago [9]. The label, for the policy head, is a one for the move played by Katago and zeros for other moves. The label, for the value head, is the evaluation of the position given by the Monte Carlo Tree Search of Katago. A value between 0 and 1 giving the probability of winning for White.

3.1   Training with Cosine Annealing

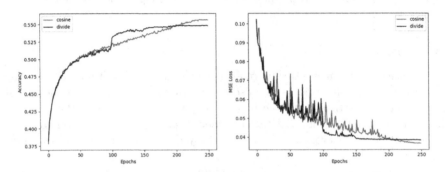

Fig. 2. Cosine annealing versus division annealing for the accuracy and the MSE of the 16 blocks mobile network. The learning rate of division annealing is divided by 10 at epoch 100, 150 and 200.

In early experiments, we tested multiple learning rate parameters using learning rate restart leads to no improvement. Figure 2 makes the comparison of cosine annealing with division annealing for the two best run. Cosine annealing ends up with better accuracy and MSE. Moreover, the learning curve for cosine annealing is smoother, for instance there are no bumps on the learning curve because of learning rate changes. So in the following experiments, we use cosine annealing without restarts, there is only one cycle.

3.2   Training Small Networks with Mixnet and Swish Activation

Figure 3 compares 16 blocks mobile networks, Mixnet and Mixnet with Swish activation. We see that Mixnet with Swish has better results than Mixnet alone, and that Mixnet alone has better result than using only 3×3 kernels. Notice that Mixnet also has similar results to a mobile network with 5×5 kernels, despite having less parameters.

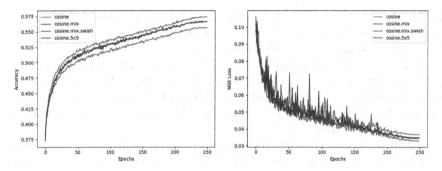

Fig. 3. Mobile network with kernels 3 × 3 and 5 × 5, Mixnet and Mixnet with Swish activation for the accuracy and the MSE of the 16 blocks mobile network.

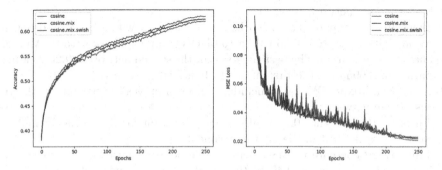

Fig. 4. Mobile network, Mixnet and Mixnet with Swish activation for the accuracy and the MSE of the 48 blocks mobile network.

3.3 Training Big Networks with Mixnet and Swish Activation

We have tested the Swish activation instead of the Rectified Linear Unit activation (ReLU) in the inverted residual blocks. Figure 4 compares 48 blocks mobile network, with 3 × 3 kernels only, Mixnet, and Mixnet with Swish activation. Mixnet with Swish activation is better than Mixnet which is better than 3 × 3 kernels only.

3.4 Playing

Table 1 gives the experiments done for comparing the different networks. Each line is the result of 400 games between two networks using given constants and numbers of playouts at each move. Each player plays 200 games as Black and 200 games as White. Both players use the Batch MCTS search algorithm [2]. In order to randomize the starting position of each game, the first 20 moves are played randomly according to the probabilities given by the policy. The properties of the Max players are given in columns 2 to 7. *Blocks* is the number of blocks of the mobile network, *Planes* is the number of planes in the trunk, M is the use of Mixnets, S is the use of the Swish activation, C is the PUCT constant and *Playouts* is the number of playouts at each move. Columns 8–13 are the properties of the Min player. The last two columns are the winrate of the Max player and the standard deviation of the winrate.

Lines 1–5 give the experiments used to find the best PUCT constant for the 16 blocks mobile network. Each constant is played against the upper constant. We observe that every constant between 0.05 and 0.30 is worse than the upper constant while 0.40 beat 0.50. We assume the best constant for 100 playouts and the 16 blocks network is 0.40.

Lines 6–9 compare the 16 blocks mobile network with Mixnet and Swish activation with a constant 0.40 to the 16 blocks mobile network with Mixnet only. Various constants are tested for the Mixnet only network and the best constant for this network is 0.20 yielding a winrate of 0.5850 for the Mixnet with Swish activation network. We can conclude that the Swish activation makes the 16 blocks mobile network stronger.

Lines 10–14 compare the 16 blocks mobile network with Mixnet and Swish activation with a constant 0.40 to the 16 blocks mobile network with no Mixnet and no Swish activation. The best constant for the second network is 0.20 yielding a winrate of 0.6950 for the first network. It is higher than the 0.5850 winrate obtained with Mixnet. We can conclude, Mixnet improves the strength of the 16 blocks mobile network.

Lines 15–16 show that 0.40 is also a good constant for the 48 blocks mobile network. We will use it in the next experiments.

Lines 17–22 compare the 48 blocks mobile network with Mixnet and Swish activation with a constant 0.40 to the 48 blocks mobile network with Mixnet only. The best performance for the second network is obtained with the 0.20 constant which gives a 0.5600 winrate for the first network. Swish activation is also beneficial to the 48 blocks mobile network.

Lines 23–27 compare the 48 blocks mobile network with Mixnet and Swish activation with a constant 0.40 to the 48 blocks mobile network without Mixnet and without Swish activation. The best performance for the second network is obtained with the 0.30 constant which gives a 0.6450 winrate for the first network. This is a worse result for the second network than with Mixnet, so Mixnet improves the strength of the 48 blocks mobile network.

Table 2 gives the accuracy, the MSE, the GPU and the CPU speed in terms of batches of size 32 processed per seconds for different networks. We did not train the residual networks on our new dataset, so we only give the speeds for residual networks [1]. In previous experiments, both the accuracy and MSE of residual networks were largely behind those of mobile networks [3]. The mobile.16.64 line is a mobile network with 16 inverted residual blocks and 64 planes in the trunk. The mobile.mix.swish.48.128 network is a mobile network with mixed convolutions, Swish activation, 48 inverted residual blocks and 128 planes in the trunk.

We can observe that the GPU speed of the networks with mixed convolutions and Swish activation is a little smaller than the speed of the original mobile networks. The accuracy and the MSE are better.

Table 1. Making the networks play.

	Blocks	Planes	M	S	C	Playouts	Blocks	Planes	M	S	C	Playouts	Winrate	σ
1	16	64	y	y	0.05	100	16	64	y	y	0.10	100	0.4250	0.025
2	16	64	y	y	0.10	100	16	64	y	y	0.20	100	0.4775	0.025
3	16	64	y	y	0.20	100	16	64	y	y	0.30	100	0.4700	0.025
4	16	64	y	y	0.30	100	16	64	y	y	0.40	100	0.4925	0.025
5	16	64	y	y	0.40	100	16	64	y	y	0.50	100	0.5025	0.025
6	16	64	y	y	0.40	100	16	64	y	n	0.40	100	0.6375	0.024
7	16	64	y	y	0.40	100	16	64	y	n	0.30	100	0.6125	0.024
8	16	64	y	y	0.40	100	16	64	y	n	0.20	100	**0.5850**	0.025
9	16	64	y	y	0.40	100	16	64	y	n	0.10	100	0.6425	0.024
10	16	64	y	y	0.40	100	16	64	n	n	0.50	100	0.8250	0.019
11	16	64	y	y	0.40	100	16	64	n	n	0.40	100	0.7475	0.022
12	16	64	y	y	0.40	100	16	64	n	n	0.30	100	0.7125	0.023
13	16	64	y	y	0.40	100	16	64	n	n	0.20	100	**0.6950**	0.023
14	16	64	y	y	0.40	100	16	64	n	n	0.10	100	0.7500	0.022
15	48	128	y	y	0.30	100	48	128	y	y	0.40	100	0.4900	0.025
16	48	128	y	y	0.40	100	48	128	y	y	0.50	100	0.5200	0.018
17	48	128	y	y	0.40	100	48	128	y	n	0.50	100	0.7125	0.023
18	48	128	y	y	0.40	100	48	128	y	n	0.40	100	0.6300	0.024
19	48	128	y	y	0.40	100	48	128	y	n	0.30	100	0.5775	0.025
20	48	128	y	y	0.40	100	48	128	y	n	0.20	100	**0.5600**	0.025
21	48	128	y	y	0.40	100	48	128	y	n	0.10	100	0.5850	0.025
22	48	128	y	y	0.40	100	48	128	y	n	0.05	100	0.6475	0.024
23	48	128	y	y	0.40	100	48	128	n	n	0.50	100	0.7050	0.023
24	48	128	y	y	0.40	100	48	128	n	n	0.40	100	0.6700	0.024
25	48	128	y	y	0.40	100	48	128	n	n	0.30	100	**0.6450**	0.024
26	48	128	y	y	0.40	100	48	128	n	n	0.20	100	0.6525	0.024
27	48	128	y	y	0.40	100	48	128	n	n	0.10	100	0.6925	0.023

Table 2. Accuracy and speed.

Network	Accuracy	MSE	GPU Speed	CPU Speed
residual.20.256			20.78	1.30
residual.40.256			12.62	0.68
mobile.16.64	55.61	0.0367	30.70	6.11
mobile.mix.16.64	56.64	0.0349	26.58	4.11
mobile.mix.swish.16.64	57.40	0.0331	21.60	2.75
mobile.48.128	61.97	0.0230	13.06	0.75
mobile.mix.48.128	62.37	0.0223	10.35	0.54
mobile.mix.swish.48.128	62.95	0.0208	7.73	0.42

4   Conclusion

We proposed three improvements to Mobile Networks for Computer Go. They improve both the supervised training and the architecture of the network by using Swish activation and mixed convolutions. The large network using mixed convolutions and the Swish activation has a winrate of 0.6450 against a similar network not using them. This brings a 104 Elo improvement. Also, the network trained with cosine annealing has better accuracy and evaluation error than the network trained dividing by 10 the learning rate.

It would be interesting to experiment these improvements in other games and also in the deep reinforcement learning framework.

Acknowledgment. This work was granted access to the HPC resources of IDRIS under the allocation 2021-AD011012539 made by GENCI.

This work was supported in part by the French government under management of "Agence Nationale de la Recherche" as part of the "Investissements d'avenir" program, reference ANR19-P3IA-0001 (PRAIRIE 3IA Institute).

References

1. Cazenave, T.: Residual networks for computer go. IEEE Trans. Games **10**(1), 107–110 (2018)
2. Cazenave, T.: Batch Monte Carlo tree search. Arxiv (2021)
3. Cazenave, T.: Improving model and search for computer Go. In: IEEE Conference on Games (2021)
4. Cazenave, T.: Mobile networks for computer Go. IEEE Trans. Games. **28**, 58–69 (2021)
5. Loshchilov, I., Hutter, F.: SGDR: stochastic gradient descent with warm restarts. arXiv preprint arXiv:1608.03983 (2016)
6. Ramachandran, P., Zoph, B., Le, Q.V.: Searching for activation functions. arXiv preprint arXiv:1710.05941 (2017)
7. Silver, D., et al.: A general reinforcement learning algorithm that masters chess, shogi, and go through self-play. Science **362**(6419), 1140–1144 (2018)
8. Tan, M., Le, Q.V.: MixConv: mixed depthwise convolutional kernels. arXiv preprint arXiv:1907.09595 (2019)
9. Wu, D.J.: Accelerating self-play learning in go. CoRR abs/1902.10565 (2019)

A Heuristic Approach to the Game of Sylver Coinage

Gilad Moskowitz[(✉)] and Vadim Ponomarenko

Mathematics Department, San Diego State University, San Diego, CA 92182, USA
gilad.moskowitz@gmail.com, vadim123@gmail.com

Abstract. Sylver Coinage is a zero-sum terminating game, making the search for an optimal strategy very enticing. Many of the challenges that existed with creating computer programs to play games like Chess and Go exist for Sylver Coinage as well. However, unlike Chess and Go, working towards finding an optimal strategy in the game of Sylver Coinage presents some new and interesting challenges. We attempt to make some headway on the problems associated with finding a strategy for Sylver Coinage using several heuristic algorithms employed by bots to play the game.

1 Introduction

John H. Conway discovered the game of Sylver Coinage, and popularized it in the 1982 book, *Winning Ways, for your mathematical plays* [1]. In this game, two players face off taking turns naming positive integers that cannot be created as a sum combination of previously named integers. The first player forced to name 1, loses. An example game is given in Table 1, below.

Since its inception, Sylver Coinage has been studied extensively in both published [3–6,9] and unpublished [2,7,8] works. A lot is known about winning and losing positions, but not much is known about how to actually win from those positions. For instance, Hutchings's Theorem [1] tells us that any prime number greater than or equal to five is a winning first move. However, there is no known strategy for finding a winning move after playing such a prime number. Another problem that comes up in the study of Sylver Coinage is that of infinite positions.

Definition 1. *An* infinite position *in the game of Sylver Coinage is one in which there are still infinitely many remaining legal moves.*

The first move played by each player is played in an infinite position, but we can see that sometimes a game can remain in an infinite position for a long time. For instance, if Player 1 plays $2^{2^{100}}$, then Player 2 can play $2^{2^{100}-1}$, then Player 1 can play $2^{2^{100}-2}$ keeping the game infinite since no odd number has been played, and this can go on for a long time. There can also be infinitely

© Springer Nature Switzerland AG 2022
C. Browne et al. (Eds.): ACG 2021, LNCS 13262, pp. 61–70, 2022.
https://doi.org/10.1007/978-3-031-11488-5_6

Table 1. Example Sylver Coinage Game.

Player	Move Played	Remaining Possible Moves	Explanation
1	3	1, 2, 4, 5, 7, 8...	Any move that is not a multiple of 3
2	5	1, 2, 4, 7	$8 = 3 + 5$, $9 = 3 + 3 + 3$, and $10 = 5 + 5$, so all moves above 7 can be made as $8 + 3k$, $9 + 3k$, or $10 + 3k$ for $k \geq 0$.
1	7	1, 2, 4	Remaining moves after playing 7.
2	4	1, 2	Remaining moves after playing 4.
1	2	1	Remaining moves after playing 2.
2	1	—	Player 2 is forced to play 1 and loses.

many possible combinations that keep a game in an infinite position. However, a game cannot remain in an infinite position forever [1]. All Sylver Coinage games will eventually end.

Comparing Sylver Coinage to games like Chess and Go, we see three distinct problems with the analysis of the game. They are the size of a position, the lack of an established corpus of human strategy, and the lack of a natural way to naively evaluate a position.

In a game of Sylver Coinage, we have the existence of infinite positions, and we already discussed how this adds difficulty to the problem. Also, many finite Sylver Coinage positions have a much larger move set than any possible position in a game of Chess or Go. This fact makes is so that computationally calculating a winning move from a given position can be almost impossible, and even a look-ahead strategy would be too computationally intensive.

Little has been discovered that concerns the strategies used to win a game of Sylver Coinage. Some positions have been analyzed completely and full winning strategies are known, similar to how there are rules to the endgame in Chess, but there are still many finite positions that have no known winning strategy. The only "opening" that is proven to be a winning strategy is a prime number greater than 5. However, there is no known strategy to winning after playing a prime number greater than 5 as an opening move. There is a complete lack of literature on attempted strategies and human evaluation of general positions. When attempting to write a bot to play a game like Chess or Go, there are many resources on strategy that can be used to improve the bot. For Sylver Coinage, no such resources exist.

Lastly, a major problem of Sylver Coinage is the lack of a naive way to evaluate a position. In Chess, for instance, pieces have value. One may evaluate positions based on material gains of each player. For instance, in most situations, a position in which a player doesn't have a queen on the board is much worse than the same position with the player still in control of their queen. In Sylver Coinage, there is no inherent way to evaluate positions and to make material gains as it were. Therefore, making a bot proves even more difficult as there is no easy implementation of a strategy that prioritizes material gains, a strategy often used historically by primitive chess bots.

With modern computational power, we hope to find new insights on the game using various algorithms that play against each other. Hence, our goals include producing a bot that plays well, and making inroads on the second and third problems. Instead of human analysts writing books on what they believe is a good position, we will have objectively skilled bots that can provide concrete data. Further, the algorithms of our best bots make progress on the third problem, by giving us a way to evaluate a position.

2   Bot Strategies for Playing Sylver Coinage

In pursuit of an optimal strategy for the game of Sylver Coinage, we began the development of several bots to play the game against each other. We will discuss the development of the heuristic strategy used by the current most successful bot and show results from testing this bot against various prior versions.

While the position is still infinite, all the bots rely on the same heuristics and some random choices to play their moves. Only once the position is finite, do the bots discussed begin to implement their strategy. For this reason, we will only focus on the strategy in a finite position. We begin by describing the first set of "naive" bots with minimal strategy. These bots were

- randomBot - Always picks a random legal move
- alwaysMin - Always picks the smallest legal move greater than 3 (since picking 1, 2, or 3 in any position will lead to a loss if the opponent plays correctly)
- alwaysMax - Always picks the largest legal move
- maximalOdd - The most complicated of the initial batch. This bot would pick the largest legal move that would return an *odd position*. If no move returns an odd position, it will just pick the largest move.

Definition 2. *We say the* parity *of a position is even when the number of remaining legal moves is even, and odd when the number of remaining legal moves is odd.*

The motivation for the maximalOdd bot came from the realization that if you can always return a position with odd parity to your opponent, you will eventually return a position where the only remaining legal moves are 1, 2, and 3 thereby winning. When we ran a round robin tournament with these four aforementioned bots, we got the results found in Table 2.

We see that the maximalOdd bot outperforms its competition. Comparing its success rate versus each individual bot over 1000 games, we found that against alwaysMax, the maximalOdd bot won all 1000 games; against alwaysMin, it won 876 games; and against randomBot, it won 760 games.

The performance of the maximalOdd bot led us to look for advancements that can be made on this strategy. We came up with the following definitions, leading to our more advanced maxThen1Weak bot.

Table 2. 4 bot Round Robin Tournament with three rounds, where each match between two bots consists of 50 games. Each match gets scored as follows, 3 points for a win, 1 point for a tie, and 0 points for a lose.

Bot Name	Match Score (27 maximum)	Total Wins	Win Percentage
maximalOdd	27	397	88.22%
alwaysMax	18	218	48.44%
randomBot	3	167	37.11%
alwaysMin	6	118	26.22%

Definition 3. *We say that a position is* weak, *or* 0-weak, *if the parity of the position is odd.*

A position in which the only remaining move is 1 is a lost position. Similarly, a position with only 1, 2, and 3 as legal moves is lost. So, we see that in many cases a position of odd parity can lead to a loss. This leads us to our next concept.

Definition 4. *We say that a position is* 1-weak *if for every move, $j > 1$ in the position, there is a different unique move, $k > 1$, such that playing k in response to j results in a weak position. Note that a 1-weak position is also a weak position as every move other than 1 has a pair that can be played to return a weak position.*

The idea here is that if Player 1 is in a 1-weak position and for every move there is a response that returns a weak position, then Player 1 will end up playing into a weak position again no matter their move. Therefore, they are unlikely to win if the opponent plays perfectly.

Working with this definition, we built a bot, maxThen1Weak, that checks all the remaining legal moves and sees if playing any of them will result in the opponent ending up in a 1-weak position. However, checking every move and then checking if a position is 1-weak is very costly in terms of computation.

Giving the bots any amount of time to calculate and play their moves could result in extremely long games. Suppose there are 10000 remaining legal moves. Testing all the moves to see if playing any of them would result in a 1-weak position is close in computational time to 10000^3 computations of position. Each computation of the remaining legal moves in a position is also costly. Thus, to prevent games from lasting too long, we implemented a time constraint on the bot to play their move. We chose to have each bot play with a 30 s chess clock, meaning that each bot has a total of 30 s for all their moves. We wanted to prevent extremely long games while still giving bots time to make calculations on critical moves.

To make sure that our maxThen1Weak bot doesn't take too long to play a move, it only starts to implement its strategy when there are less than thirty remaining legal moves. Until then it will play the maximal legal move remaining. Even with this restriction on the implementation of the strategy, the bot still vastly outperforms the competition thus far. In a round robin tournament against all the previously mentioned bots, the maxThen1Weak bot performed as seen in Table 3.

Table 3. 5 bot Round Robin Tournament with three rounds, where each match between two bots consists of 50 games. Each match gets scored as follows, 3 points for a win, 1 point for a tie, and 0 points for a lose.

Bot Name	Match Score (36 maximum)	Total Wins	Win Percentage
maxThen1Weak	36	585	97.5%
alwaysMax	18	213	35.5%
alwaysMin	9	135	22.5%
maximalOdd	27	418	69.67%
randomBot	0	149	24.83%

We can see that the maxThen1Weak bot was much ahead of its competition, only losing 15 out of the 600 games played. In a head-to-head match of 1000 games against the maximalOdd bot, the maxThen1Weak bot was able to win 934 games.

However, as it turns out, there are many positions that are 1-weak, but not lost. To address this, we arrive at our next definition.

Definition 5. *We say that a position is* 2-weak *if for each move, $j > 1$ in the position, there is a different unique move, $k > 1$, such that playing k in response to j returns a 1-weak position.*

Remark 1. When checking if a position is 2-weak, we are seeing what the position will be after four moves have been played. In many finite positions, we can try to calculate every possible remaining position to find a complete strategy for winning, but this is very computationally expensive. With the strategy of 2-weak we only need to look a few moves ahead to give us a decent sense of the position.

So, using this new definition, we created the maxThen2Weak bot (previously called strongCounterV2, winner of the 2021 Computer Olympiad). This bot, like the maxThen1Weak bot, only uses its strategy when there are less than thirty remaining legal moves. With this modification, we saw significant improvement. When we added maxThen2Weak to the round robin tournament we got the results in Table 4.

We see that the maxThen2Weak bot does have a slight edge over the maxThen1Weak bot, but both are much stronger than the rest of the bots. In a head-to-head match of 1000 games against the maxThen1Weak bot, the maxThen2Weak bot was able to win 664 games.

Using the notion of a 2-weak position we devised an improvement for the maxThen2Weak bot. The pseudo-code for this bot, dubbed peekThen2Weak (previously called strongCounterV3), follows.

Table 4. 6 bot Round Robin Tournament with three rounds, where each match between two bots consists of 50 games. Each match gets scored as follows, 3 points for a win, 1 point for a tie, and 0 points for a lose.

Bot Name	Match Score (45 maximum)	Total Wins	Win Percentage
maxThen2Weak	45	674	89.87%
maxThen1Weak	36	650	86.67%
maximalOdd	27	423	56.4%
alwaysMax	15	216	28.8%
alwaysMin	6	113	15.07%
randomBot	6	174	23.2%

```
def pretendMove(move, remainingMoves):
    return [i if(i in remainingMoves after move is played)]

if(numberOfRemainingMoves <30):
    for i in remainingMoves:
        nextPosition = pretendMove(i, remainingMoves)
        check if nextPosition is 2-weak
        if yes:
            play i
    if no i returns a 2-weak position:
        play max remainingMove

else if(numberOfRemainingMoves <200):
    for i in remainingMoves:
        nextPosition = pretendMove(i, remainingMoves)
        if(numberOfMovesInNextPosition <20):
            check if nextPosition is 2-weak
            if yes:
                play i
    if no i returns a 2-weak position:
        play max remainingMove

else:
    play max remainingMove
```

The logic of the bot is to see if playing a move will return a 2-weak position, thereby giving the opponent a bad position to play from. If there is a move that returns a small 2-weak position, play it. If not, play the largest remaining move, which will only get rid of one move, and hope that your opponent makes a mistake, letting you put them in a 2-weak position. One of the big goals for this bot is to make it play efficiently and keep the calculation times to a minimum. This means that we can't test for 2-weak positions after playing every move in a large position since the calculation time for that would be close to n^5 where

n is the number of remaining moves. Through a lot of testing, we found that starting the checks for 2-weak positions when the number of remaining moves was less than 30 was optimal for maximizing the number of wins while staying within the time constraint. Although this is a relatively small position, there are still many positions that fall into this category. We have also seen that in larger positions opposing bots rarely have a better strategy and therefore often lose once the position becomes small.

Adding this bot to the round robin we have been running, we got the following results (Table 5).

Table 5. 7 bot Round Robin Tournament with three rounds, where each match between two bots consists of 50 games. Each match gets scored as follows, 3 points for a win, 1 point for a tie, and 0 points for a lose.

Bot Name	Match Score (54 maximum)	Total Wins	Win Percentage
peekThen2Weak	52	781	86.78%
maxThen2Weak	43	722	80.22%
maxThen1Weak	39	707	78.56%
maximalOdd	27	427	47.44%
alwaysMax	16	221	24.56%
alwaysMin	6	122	13.56%
randomBot	4	170	18.89%

3   Introducing Elo to Keep Track of Bot Success

The Elo rating system, named after Arpad Elo, is commonly used to represent the relative skill level of players in various games including board games like chess and Scrabble, and video games like League of Legends. Even though the player pool was small, the results of the round robin tournaments showed that certain bots are almost always able to beat other bots. This means that win percentage cannot be used as a good metric to determine the skill level of the bot since it can vary based on the competition. This led to the idea of implementing an Elo ranking system for the bots. That way, bots could be tiered in some capacity and we can get a sense of how likely one bot is to beat another. An Elo ranking is also future-proof. As more competitions are held and more bots compete, an Elo ranking allows for more accurate comparison of bots, even ones that did not play in the same tournament.

To generate an Elo ranking for each of the bots, we used a series of round robin tournament with Elo for each bot being calculated after it completed a match. Suppose bot A is playing against bot B. After the match the winning bot gets a score of 3 and the losing bot gets a score of 0. In the case of a tie, both bots get a score of 1. The formula for calculating each bots Elo is

$$E_A = \frac{1}{1 + 10^{\frac{R_B - R_A}{400}}}$$

$$E_B = \frac{1}{1 + 10^{\frac{R_A - R_B}{400}}}$$

$$R'_A = R_A + 20 * (BotAScore - 2 * E_A)$$

$$R'_B = R_B + 20 * (BotBscore - 2 * E_B)$$

where R_A and R_B are the bots' Elos going into the match and R'_A and R'_B are the bots' Elos following the match. The scoring was set in this manner so that if two bots of equal Elo played and the match resulted in a tie, neither bot's Elo would change. Also, with this system the bot's Elo would change aggressively based on a win or loss. This was done to make sure that we can get fairly accurate Elos quickly. All the bots were given a base Elo of 800 except for alwaysMin with an Elo of 400, maxThen1Weak with an Elo of 1600, and a few other bots with various strategies. We then ran a round robin tournament with 14 bots including all the ones mentioned so far and one AI that was trained via Deep Reinforcement Learning. The tournament consisted of 4 rounds with each match having 25 games, with peekThen2Weak coming in first and maxThen2Weak as the runner-up. We set the bots Elos in accordance with the results of this round robin and ran another round robin with the same bots. This time, three rounds with each match having 30 games, again we ended with peekThen2Weak coming in first and maxThen2Weak as the runner-up. After this, we scaled all the bots Elos by setting the Elo of the worst bot, alwaysMin, to 400 and adjusting the rest of the Elos accordingly.

At this point, we had the following Elo ratings for each of the aforementioned bots (Table 6).

Table 6. Bots and their respective Elo ratings after a series of round robin tournamets to initialize them.

Bot Name	Elo
peekThen2Weak	2605
maxThen2Weak	2514
maxThen1Weak	2279
maximalOdd	1313
alwaysMax	759
randomBot	429
alwaysMin	400

Once we had a fairly accurate Elo rating for each bot, we changed the formula for calculating Elo. Now, a winning bot would get a score of 1, a losing bot would get a score of 0, and in the case of a tie each bot would get a score of 0.5. The new Elo formula is

$$E_A = \frac{1}{1 + 10^{\frac{R_B - R_A}{400}}}$$

$$E_B = \frac{1}{1 + 10^{\frac{R_A - R_B}{400}}}$$

$$R'_A = R_A + 16 * (BotAScore - 2 * E_A)$$

$$R'_B = R_B + 16 * (BotBScore - 2 * E_B).$$

This is a more conservative system so that the bots don't change score so drastically. We hope that as more bots and algorithms are developed, we can use this Elo system to determine their success more accurately.

4 Conclusion

One of the problems we mentioned earlier with the journey to finding an optimal strategy for the game of Sylver Coinage is the lack of a way to evaluate a position. Through our research, we've developed four ways of analyzing a position 0-weak, 1-weak, 2-weak, peek-then-2-weak. Building on this, we can discuss two more definitions.

Definition 6. *We say that a position is n-weak if for each move, $j > 1$ in the position, there is a different unique move, $k > 1$, such that playing k in response to j returns a (n - 1)-weak position. For example, in a 3-weak position, for every $j > 1$ there is a response $k > 1$ such that the resulting position after playing both moves is a 2-weak position.*

Definition 7. *We stay that a position is ∞-weak if for every move greater than 1 that the current player can play for the rest of the game, their opponent will always have a unique response greater than 1 that returns a weak position.*

We see that an ∞-weak position is a lost position as the opponent will always have a response to keep the parity odd, eventually returning the position with 1 being the only legal move. It is important to note that finding an ∞-weak position is not guaranteed, and that not all lost positions are necessarily ∞-weak. Also, it is very computationally difficult to determine if a position is ∞-weak as we have to calculate every possible remaining position. Even just calculating if a position is n-weak for $n > 2$ gets very computationally expensive very quickly. However, surely finding a move that puts our opponent in a 3-weak position is better than a move that puts our opponent in a 2-weak position. Therefore, logic says that with enough computational power, a bot that searches for a move that returns a 3-weak position, then, failing to find one, searches for a move that returns a 2-weak position, and so forth would be stronger than our current bots. This shows that still, there is much work to be done in the search for an optimal strategy for the game Sylver Coinage.

Our bots provide headway in addressing the second and third problems we mentioned in the introduction regarding the analysis of the game. With regards to the second problem, the various strategies implemented by the bots provide some material concerning strategies for winning a game of Sylver Coinage. With regards to the third problem, although we still have no way of naively evaluating a position, the hierarchical nature of our bot strategies suggests that our analysis of position has some value. That leads to the likelihood of finding ways to evaluate a position, even if that value comes from a bot's analysis.

We hope that with the development of the bots thus far and the growth of Sylver Coinage as a competitive bot played game, we can make great headway on the problems associated with finding a winning strategy for the game.

References

1. Berlekamp, E.R., Conway, J.H., Guy, R.K.: Winning Ways, for Your Mathematical Plays. Academic Press, London (1982)
2. Blok, T.: Sylver coinage positions with g=2 (2021). https://userpages.monmouth.com/~colonel/sylver/Sylver_Coinage_positions_with_g=2.pdf
3. Eaton, R., Herzinger, K., Pierce, I., Thompson, J.: Numerical semigroups and the game of Sylver Coinage. Am. Math. Monthly **127**, 8 (2020)
4. Guy, R.K.: Twenty questions concerning conway's sylver coinage. Am. Math. Monthly **83** (1976)
5. Michael, T.S.: How to Guard an Art Gallery and Other Discrete Mathematical Adventures. Johns Hopkins University Press, Baltimore (2009)
6. Nowakowski, R.J.: ..., Welter's Game, Sylver Coinage, Dots-and-Boxes, ... In: Combinatorial Games (Columbus, OH, 1990), In: Proceedings of Symposia in Applied Mathematics, vol. 43, pp. 155–182. American Mathematical Society, Providence, RI (1991). https://doi.org/10.1090/psapm/043/1095544
7. Sicherman, G.: New results in Sylver Ccoinage (1991), https://userpages.monmouth.com/~colonel/sylver/index.html
8. Sicherman, G.: Late news of Sylver Coinage (1996), https://userpages.monmouth.com/~colonel/sylver/index.html
9. Sicherman, G.: Theory and practice of sylver coinage. Integers **2** (2002)

Evaluating Interpretability Methods for DNNs in Game-Playing Agents

Aðalsteinn Pálsson$^{(\boxtimes)}$ and Yngvi Björnsson(iD)

Department of Computer Science, Reykjavik University, Reykjavik, Iceland
adalsteinn19@ru.is

Abstract. There is a trend in game-playing agents to move towards an Alpha-Zero-style architecture, including using a deep neural network as a model for evaluating game positions. Model interpretability in such agents is problematic. We evaluate the applicability and effectiveness of several saliency-map-based methods for improving the interpretability of a deep neural network trained for evaluating game positions, using the game of Breakthrough as our testbed. We show that the more applicable methods provide insights into the importance of the different game pieces and other domain-dependent knowledge learned by the model.

Keywords: game-playing · model-interpretability · deep neural-networks

1 Introduction

Over the past few decades, research into game-playing programs for abstract strategy board games has first-and-foremost concentrated on developing new techniques and algorithms for improving gameplay. That work has resulted in super-human strength game-playing agents [9] for disparate games such as chess, checkers, Othello, Go, and many more. At the same time, other important aspects of intelligent systems have been mainly neglected, such as how to explain the rationality for one's actions in human-understandable terms. Moreover, recent advancements in the field where game-playing agents use deep neural networks (DNNs) to evaluate board positions and action selection in the think-ahead process make the decision-making process even more non-transparent.

The above-mentioned lack of transparency is not specific to game-playing agents. As computer-generated models in disparate fields such as healthcare and banking have become increasingly ubiquitous, the need for humans to understand their decisions becomes increasingly crucial for establishing trustworthiness. This has spurred research interest in *model interpretability*, that is, the development of approaches to make it less complicated for humans to understand the cause of models' decisions in terms of their inputs. Several such approaches now exist, for example, in the field of image recognition, which has hitherto been at the forefront of both deep neural network and model-interpretability research. In

© Springer Nature Switzerland AG 2022
C. Browne et al. (Eds.): ACG 2021, LNCS 13262, pp. 71–81, 2022.
https://doi.org/10.1007/978-3-031-11488-5_7

particular, *saliency maps* [10] have become a popular way of visualizing which regions of an image are primarily responsible for a given classification decision.

In this work, we evaluate the applicability and effectiveness of several saliency-map-based methods for improving the interpretability of a neural network trained to evaluate game positions of a board game, using the game Breakthrough as our testbed. The paper's primary contributions are: (i) We evaluate several popular saliency-map-based methods within recently established paradigms and taxonomies for black-box interpretability methods; and (ii) show how they can be applied to interpret a deep neural network model for an abstract board-game—a domain they are not mainly intended for; and, finally; (iii) assiduously evaluate and rank the methods by their effectiveness in our domain. Furthermore, this works adds to the recently emerging literature on explaining models and actions learned by game-playing agents [4,6].

The paper is organized as follows. Section 2 introduces the terminology and preliminaries. Section 3 explains the game-playing agent, model, and evaluation methods used. In Sect. 4, which constitutes the main body of the work, we introduce and analyze the finding of the empirical evaluations of the saliency-maps methods. Finally, in Sect. 5, we conclude and discuss future work.

2   Background

We start with a high-level overview of the model-interpretability methods we investigate, before explaining the rules of the game of Breakthrough.

2.1   Model Interpretability

The taxonomy of explanation methods of black-box models categorizes them as either *global* or *local* and *model-specific* or *model-agnostic*. Global methods create explanations valid across all input instances, while local methods' explanations are specific to individual instances. Model-agnostic methods explain any black-box models, while model-specific methods leverage the model's architecture.

The most straightforward local method is *occlusion*, where the model's output sensitivity to leaving out (zeroing) arbitrary input parameters is investigated [15]. Such an approach, where applicable, is appealing as it is both model-agnostic and straightforward to implement.

A method that is local and model-specific but still requires no ad-hoc work is to analyze the gradient of the output with respect to individual input pixels [10]. Multiplying the input with the gradient is also often preferable because it leverages the strength of the input. A further extension on the gradient method is Integrated Gradients [12], which relies on a baseline and interprets the input feature attribution as the integration of gradients on the straight-line path between the input and the baseline. GradientShap [5] is an extension of Integrated Gradients that computes the expected gradient by sampling baseline values.

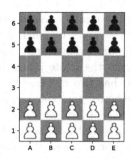 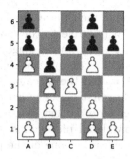

Fig. 1. Breakthrough: Initial board position (left); Example position (right)

DeepLIFT [8] is a model-specific explanation method that does not rely on the gradient. It overcomes the limitations of loss of information when the gradient is zero because the signal might still be meaningful. It calculates the importance in a backward fashion by distributing attributions, or blame, in terms of difference-from-reference. For all neurons, a difference-from-reference is calculated by passing through the input sample and the reference. Finally, it calculates the importance using predefined rules, such as the linear- or reveal cancel rule.[1]

LIME [7] is a local model-agnostic method, which uses an interpretable surrogate model to explain the black-box model. A local model, such as a linear model, is trained on a dataset derived from sampling noise around the input and using the model evaluation as a target. In the case of a linear model, the weights from the model serve as feature attributions.

The Shapley value [11] is a concept from cooperative game theory that can be used to calculate feature attribution. There are multiple ways to approximate the Shapley values [5], the one we use in this paper is Shapley Value Sampling. It takes random permutations of the input and adds them one by one to the baseline. This is repeated multiple times to approximate the Shapley values.

2.2 Breakthrough

The game Breakthrough is an abstract strategy board game, originally played on a 7×7 board but later popularized to an 8×8 board. The game can be played on different-sized (not necessarily squared) boards. In this work we use a smaller variant of 5×6, mainly for the ease of demonstration.

The game is a two-player turn-taking game. The players are referred to as White and Black, respectively. The board is initially set up by placing White's pieces along the first two rows and the Black pieces on the last two rows, as shown in Fig. 1 (left). White goes first and then players alternate, with each

---

[1] A baseline, or a reference, may be interpreted as a neutral state of a neural network, and is important for defining counterfactual arguments [12]. When assigning an attribution/blame to the input, it is done relative to the baseline. Most of the methods we consider in this paper rely on a baseline defined as all-zeros.

player moving one of their pieces per turn. A piece moves one square straight or diagonally forward (relative to the player). A straight forward move is allowed to an empty square only, but a diagonally forward moves may also capture an opponent's piece. For example, in Fig. 1 (right) the white pawn on $d2$ has two moves, to $d3$ or $e3$, and the piece on $d4$ also has two moves, to $c5$ or $e5$, both with a capture. The first player to get a piece across the board wins: White wins by moving a piece onto the last row, and likewise, Black wins by moving a piece onto the first row. If all pieces of a player are captured, that player loses. It follows from the rules that one of the players always wins (there are no draws).

3   Methods

An Alpha-Zero-like agent for playing the game of Breakthrough was developed for the paper.[2] The following subsection providing details, whereas the next subsection gives an overview of our evaluation methodology.

3.1   The Model

We trained a AlphaZero-like model [9], $(p, v) = f_\theta(s)$ with parameters θ, where p is the policy and v is the value function. The value is the output from a tanh activation function, a scalar value between -1 and 1. And the policy is a tensor with three channels, where each channel is the same size as the board and encodes the three different move directions available for all the pieces, i.e., forward, and diagonally left, or -right.

The model consists of a body of 5 residual blocks with 56 filters, followed by the policy and value heads. The input to the model is a tensor with three channels, where each channel is the same size as the board. The first channel encodes the board positions of the active player (the player to move), the second channel encodes the positions of the opponent, and the third channel is a binary encoding of the active player's color. If it is white to move, then the third channel is all ones. Otherwise, it is all zeros. In our case, we use a board with six rows and five columns. The search is performed with the Monte Carlo tree search algorithm like AlphaZero. The training procedure was via asynchronous self-play, where each move played used 200 simulations.

We deviate from AlphaZero we only feed the model the current board position. We do this mainly to make the model easier to explain, as an explanation should not depend on previous board positions.

3.2   Evaluation Measures

The goal of this paper is to compare and evaluate model explanation methods in the domain of game-playing. Our model is returning a value estimate, and

---

[2] We have no objective measure of the agent's playing strength, but as anecdote, it consistently wins all humans it plays, including an expert-level chess player.

the goal is to explain the estimation. When evaluating the usefulness of the explanations, we will consider if we can use the explanations to gain trust in the trained model, if the explanation satisfies our human curiosity, and if we can interpret some meaning from the model [1].

The evaluation approaches are split into three categories [3]: (i) Functionally-grounded, where the explanation is evaluated without human, using a proxy as an indication of explainability; (ii) human-grounded, requiring a human with non-expertise to evaluate a simple explanation and; (iii) application-grounded, which requires a human-expert, evaluating an explanation for a real-world task.

At first, we will inspect the saliency map from a qualitative human-grounded perspective. We will briefly debate if the saliency maps match our human objectives by visualizing statistics from the saliency methods.

Quantitative evaluation will be in the form of functionally grounded experiments. We will define three tasks where the performance will be used as a proxy for explanation quality. First, we will assess if the saliency method assigns the highest saliency to a critical piece, and conversely, if the lowest saliency is assigned to a non-critical piece. We will analyze this as an ablation study, where we separately remove the least and most important piece and measure its impact on the game's outcome. The second proxy task is to analyze the saliency of the piece that ultimately secures the win in a self-play game. The third task is to find the smallest sufficient subset of pieces required to retain a winning position. Then we iteratively remove non-important pieces according to an explainability method. The explainability method that has the highest area under the curve has the highest explainability quality.

4   Results

We ran several experiments, both for contrasting the effectiveness of different model-interpretability methods and for gaining added insights into the domain-dependent knowledge captured by the learned model.

4.1   Experimental Setup

The model is implemented in Pytorch and trained using asynchronous self-play using two GeForce RTX 3090, AMD Ryzen 9 5950 with 64 GB of RAM, and using RAY [14]. It was trained while playing a total of 630,000 self-play games. The training used stochastic gradient descent with a batch size of 512 and a decaying learning rate that was re-initialized every few thousand iterations.

LIME's perturbation of the input was binary, i.e., the features were set either to zero or one, allowing a more direct comparison with the occlusion method.

The saliency map methods in the paper used the implementations in the model interpretability library Captum [13].

4.2   Saliency Methods: Qualitative Evaluation

In image classification, a saliency map is a two-dimensional image, representing each pixel's perceived importance for the model's output. The analogy for a board game would be an image of the board's squares, representing the square's (or the piece on it) influence on the model's evaluation of the current game state.

Model-agnostic methods typically rely on modifying the input somehow and observing its effect on the model's output. That way, each input feature's attribution (or importance) to the output can be determined. One of the most straightforward of such ablation methods is that of *occlusion*, where in our domain we remove a piece from the board and observe the effect of the model's output. Figure 2 shows the saliency map from such an experiment.[3] It shows clearly that the attacking piece on $d4$ is White's primary asset along with the supporting piece on $c3$ (and the independently potential breakthrough piece on $a4$). Unsurprisingly, for Black, the defending pieces on $a6$, $a5$, $d6$ and $d5$ play the most crucial role. This assessment is in perfect consonance with human (expert-level) assessment: the white pieces on $d4$ and $c3$, with White to move, can collectively win the game on their own, while the piece on $a4$ is a valuable long-term asset severely restricting the mobility of two of the black pieces, thus potentially placing Black later in *zugzwang*, but a well-established expert-level strategy in playing Breakthrough is to force such situations.

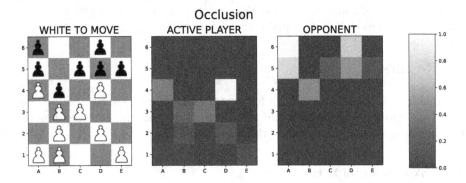

Fig. 2. The saliency map generated by Occlusion; the lighter a square is, the more important the piece occupying the square is.

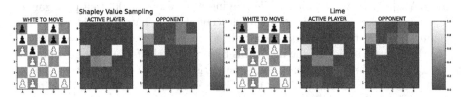

Fig. 3. Model-agnostic saliency methods: Shapley Value Sampling and Lime

---

[3] For ease of comparison, the maps are scaled to be in the range [0.0–1.0].

We ran the same type of experiment for two additional model-agnostic algorithms, *Shapley Value Sampling (SVS)* and *LIME*, which also find attribution by perturbing the model input parameters. However, they do it in a more refined way, potentially detecting non-trivial input interactions (as described in Sect. 2). The two methods give almost identical results, depicted in Fig. 3, with the result for the most part also consistent with the one from the occlusion method. The only significant difference is that for Black, the defending player, the piece on b4 gets much-added importance (without that defending piece, White will have additional ways to win by immediately playing a piece to that square).

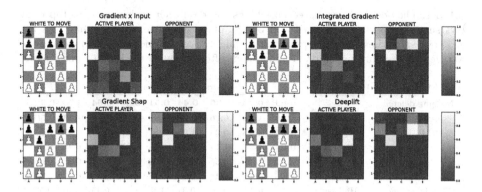

Fig. 4. Model-specific saliency methods.

We furthermore experimented with a few model-specific approaches as taking advantage of the model's internals may (in theory) yield further benefits. We looked at several methods that use the model's gradient in different ways to detect attribution, and one non-gradient-based, Deeplift. Figure 4 shows the resulting saliency maps. Again, the more sophisticated methods, Integrated Gradient, GradientShap, and Deeplift, all give intuitively plausible results, whereas the straightforward gradient-based method is more indecisive.

4.3 Saliency Methods: Quantitative Evaluation

Qualitative evaluation as in the previous subsection, albeit able to provide valuable insights, is not sufficient for determining the relative effectiveness of the different saliency map algorithms—a quantitative approach is needed for that.

We generated 10,000 game positions from self-play by stopping play randomly 10–30 moves (plies) into the game. We played two games from each position for each saliency-based approach: one without intervention and one after removing the most important piece for the player to move as judged by the respective saliency method. The expectation is that the more reliable indicators of a most-importance piece suffer more profound drop in proportion of games won.

Table 1 summarizes the results. We see the expected effect in all cases but most profoundly for the *Occlusion* method followed closely by *LIME*, *SVS* and

DeepLift. This gives us added confidence that these saliency methods are detecting the importance of the different pieces.

In an attempt to further discriminate the effectiveness of the different approaches in detecting valuable pieces, we looked at how important a pawn reaching the opponent's back rank was judged a few moves earlier. One can think of that information as an indicator of how quickly a particular saliency method realizes the importance of such "breakaway" pieces. Figure 5 shows that information, and apparently, *LIME* and *SVS* seem to put much-added importance on such pieces, whereas *Occlusion* and *Gradient* do not.

Table 1. The positions are placed into 9 bins based on their evaluation. The top-most rows shows proportion of games won by the player to move, for each bin, and the next rows show the same after removing the highest ranked piece. We also include the average and standard deviation of all methods after removing the lowest ranked piece.

Method	Importance	Proportion of games won								
Nothing deleted	-	0.09	0.26	0.36	0.44	0.51	0.55	0.64	0.74	0.90
Occlusion	Highest	0.08	0.21	0.26	**0.29**	0.31	0.37	0.40	0.43	0.50
LIME	Highest	**0.04**	0.18	0.26	**0.29**	0.34	0.39	0.45	0.47	0.51
SVS	Highest	**0.04**	0.19	**0.24**	0.32	0.38	**0.37**	0.44	0.45	0.51
Gradient	Highest	0.12	0.23	0.35	0.41	0.49	0.50	0.57	0.63	0.66
Integ. Gradients	Highest	0.06	0.18	**0.24**	0.32	0.38	0.41	0.44	0.47	0.56
Gradient Shap	Highest	0.05	**0.15**	0.27	0.31	0.36	0.41	0.45	0.49	0.57
Deeplift	Highest	**0.04**	0.18	**0.24**	0.34	0.35	0.40	0.46	0.47	0.51
Average	Lowest	0.10	0.27	0.37	0.44	0.50	0.56	0.62	0.75	0.91
Std Dev	Lowest	0.01	0.03	0.05	0.06	0.06	0.05	0.04	0.02	0.00
Mean bin value (before deletion)		**-0.90**	**-0.68**	**-0.45**	**-0.22**	**0.00**	**0.22**	**0.45**	**0.67**	**0.90**

It is also of interest to investigate how confidently the methods rank the less important pieces. For that, we use the (functionally grounded) method of smallest sufficient subsets [2], which in our domain translates into the set of pieces required to retain a winning position. To create a test-suite, we sampled positions from random games according to the model's value function to find positions in the current player being only a slight favorite. Then we gradually removed the pawns considered least important, one at a time, and recorded its effect on the proportion of games won. Figure 6 depicts the result. Essentially, the later a curve drops, the more effective the respective saliency method is in ranking pieces by importance. There are two clear winners, *LIME* and *SVS*, and two methods that do notably worse than the others, *Occlusion* and *Gradient*.

4.4   Explaining the Explanations

Finally, we were also interested in knowing common higher-level characteristics of pieces judged valuable. One way to unveil such characteristics is to build a surrogate model from hand-made higher-level features and then train the surrogate model to predict the saliency values.

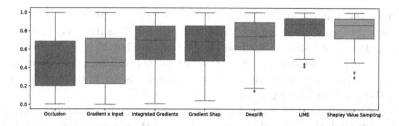

Fig. 5. Distribution of saliency values of a piece 4-ply prior to the winning move.

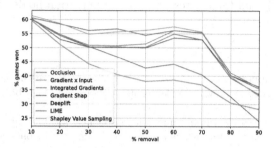

Fig. 6. Effect of gradually removing the least-important pieces.

We build such a surrogate model using a LightGBM regression tree. Figure 7 shows the relative importance of higher-level features we defined for the model; it clearly shows how important it is in Breakthrough to have advanced pieces, but features such as a center-of-mass are also important.

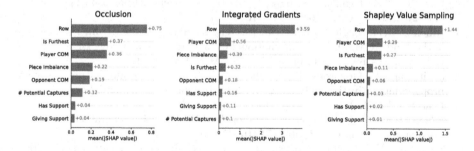

Fig. 7. The average Shapley Values of each of the inputs to the interpretable surrogate model. Here COM stands for center of mass. Has- and Giving Support indicate if the piece is supporting or giving support diagonally to a piece of the same color. # Potential Captures indicates is the number of available capture moves for the current player.

5   Conclusion and Future Work

This paper evaluated several popular saliency-based model interpretability methods on a DNN based game-playing agent, demonstrating their usefulness for identifying the most and least essential game pieces. The more sophisticated attribution methods, like Shapley Value Sampling and LIME, performed overall the best. One of the strengths of those methods is that they can capture non-trivial interactions between the inputs, which seems well suited to identify various in-game piece dynamics. Moreover, those methods are both model-agnostic, making them well-suited for a wide range of models.

As future work, we plan to evaluate the methods' applicability in other games to better establish their usefulness in the domain of abstract board games. Also, we only scratched the surface of looking at higher-level domain concepts (beyond piece importance), and further research in that direction holds promise.

References

1. Bodria, F., Giannotti, F., Guidotti, R., Naretto, F., Pedreschi, D., Rinzivillo, S.: Benchmarking and survey of explanation methods for black box models. CoRR, abs/2102.13076 (2021)
2. Dabkowski, P., Gal, Y.: Real time image saliency for black box classifiers. In: Advances in Neural Information Processing Systems, pp. 6968–6977, December 2017
3. Doshi-Velez, F., Kim, B.: Towards a rigorous science of interpretable machine learning. arXiv:1702.08608 (2017)
4. Fritz, J., Fürnkranz, J.: Some chess-specific improvements for perturbation-based saliency maps. In: IEEE Conference on Games (CIG) 2021, Copenhagen, Denmark. IEEE (2021)
5. Lundberg, S.M., Lee, S.I.: A unified approach to interpreting model predictions. In: Advances in Neural Information Processing Systems, pp. 4766–4775, December 2017
6. Puri, N., et al.: Explain your move: understanding agent actions using specific and relevant feature attribution. In: ICLR (2020)
7. Ribeiro, M.T., Singh, S., Guestrin, C.: Why should i trust you?. In: Proceedings of the 22nd ACM SIGKDD International Conference on Knowledge Discovery and Data Mining - KDD 2016, pp. 1135–1144. ACM Press (2016)
8. Shrikumar, A., Greenside, P., Kundaje, A.: Learning important features through propagating activation differences. In: 34th International Conference on Machine Learning, ICML 2017, vol. 7, pp. 4844–4866 (2017)
9. Silver, D., et al.: Mastering chess and shogi by self-play with a general reinforcement learning algorithm. CoRR, abs/1712.01815 (2017)
10. Simonyan, K., Vedaldi, A., Zisserman, A.: Deep inside convolutional networks: visualising image classification models and saliency maps. CoRR, abs/1312.6034 (2014)
11. Štrumbelj, E., Kononenko, I.: Explaining prediction models and individual predictions with feature contributions. Knowl. Inf. Syst. **41**(3), 647–665 (2013). https://doi.org/10.1007/s10115-013-0679-x
12. Sundararajan, M., Taly, A., Yan, Q.: Axiomatic attribution for deep networks. In: 34th International Conference on Machine Learning, ICML 2017, vol. 7, pp. 5109–5118 (2017)
13. The PyTorch Team: Model interpretability and understanding for PyTorch (2021). https://captum.ai. Accessed 20 Sept 2021

14. The Ray Team: Ray provides a simple, universal API for building distributed applications (2021). https://docs.ray.io. Accessed 20 Sept 2021
15. Zeiler, M.D., Fergus, R.: Visualizing and understanding convolutional networks. In: Fleet, D., Pajdla, T., Schiele, B., Tuytelaars, T. (eds.) ECCV 2014. LNCS, vol. 8689, pp. 818–833. Springer, Cham (2014). https://doi.org/10.1007/978-3-319-10590-1_53

Solving Games

Quixo is Solved

Satoshi Tanaka[1], François Bonnet[1(✉)], Sébastien Tixeuil[2],
and Yasumasa Tamura[1]

[1] Tokyo Institute of Technology, Tokyo, Japan
tanaka.s@coord.c.titech.ac.jp, {bonnet,tamura}@c.titech.ac.jp
[2] Sorbonne Université, CNRS, LIP6, Paris, France
Sebastien.Tixeuil@lip6.fr

Abstract. Quixo is a two-player game played on a 5×5 grid where
the players try to align five identical symbols. Using a combination of
value iteration and backward induction, we propose the first complete
analysis of the game. We describe memory-efficient data structures and
algorithmic optimizations that make the game solvable within reasonable
time and space constraints. Our main conclusion is that Quixo is a Draw
game. The paper also contains the analysis of smaller boards and presents
some interesting states extracted from our computations.

Keywords: Quixo · Strongly Solved · Backward Induction · Draw
Game

1 Introduction

1.1 Quixo

Quixo is an abstract strategy game designed by Thierry Chapeau in 1995 and
published by Gigamic [3,4]. Quixo won multiple awards,[1] both in France and
in the United States. While a four-player variant exists, Quixo is mostly a two-
player game that is played on a 5×5 grid, also called *board*. Each grid cell, also
called *tile*, can be empty, or marked by the symbol of one player: X or O.

At each turn, the active player first *(i)* takes a tile – empty or with her symbol
– from the border (*i.e.* excluding the 9 central tiles), and then *(ii)* inserts it, with
her symbol, back into to the grid by pushing existing tiles toward the hole created
in step *(i)*. The winning player is the first to create a line of 5 tiles all with her
symbol, horizontally, vertically, or diagonally. Note that if a player creates two
lines with distinct symbols in a single turn, then the opponent is the winner.

---

[1] As d'Or Festival International des Jeux (1995), Oscar du Jouet (1995), Mensa Select
Top 5 Best Games (1995), Games Magazine "Games 100 Selection" (1995), Games
Magazine "Best New Strategy Game" (1995), Parent's Choice Gold Award (1995) [4].

A Japanese version of this paper was published and presented at the 25th Game
Programming Workshop (GPW'20) [6]. S. Tanaka received the GPW Research
Encouragement Award and the IPSJ Yamashita SIG Research Award for his work.

C. Browne et al. (Eds.): ACG 2021, LNCS 13262, pp. 85–95, 2022.
https://doi.org/10.1007/978-3-031-11488-5_8

Figures 1a and 1b show the real game and our corresponding representation. Figure 1c depicts the resulting board after a valid turn by player O from the board depicted in Fig. 1b: player O first *(i)* takes the rightmost (empty) tile of the second row, and then *(ii)* inserts it at the leftmost position shifting the other tiles of this second row to the right.

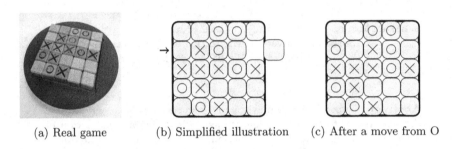

(a) Real game   (b) Simplified illustration   (c) After a move from O

Fig. 1. The two-player game of Quixo

Quixo bears an immediate resemblance with some classic games such as Tic-Tac-Toe, Connect-Four, or Gomoku. However there are two major differences: *(1)* the board is "dynamic;" a placed X or O may change its location in subsequent turns, *(2)* the game is unbounded (in term of turns). The first point is what makes the game interesting to play: dynamicity makes Quixo very difficult for a human player to plan more than a couple of turns in advance. The second point raises immediately a natural question about termination. Trivially, players could "cooperate" to create an infinite game, as official rules do not specify any terminating rules, such as the 50-move or the threefold repetition rules of chess.

1.2   Objectives and Challenges

Our main goal is to solve Quixo, which means finding the optimal outcome of the game assuming perfect players. As for any combinatorial game, there are only three possible outcomes: first-player-Win, second-player-Win, or Draw. While a second-player-Win seems unlikely, there is no easy observation (e.g. strategy stealing argument) that would permit to discard it. It is improbable to obtain analytical results, so we focus on computing this optimal outcome and the corresponding optimal strategies. More precisely, we are looking for outcomes of all states, i.e. strongly-solving the game. In addition to the real 5×5 game of Quixo, we also analyze variants using 3×3 or 4×4 grids, where players have to create a line of 3 (resp, 4) tiles with their symbols.

Even on the 5×5 grid, Quixo's game "tree" is not extremely large, in comparison with other games. The number of positions is upper bounded by $2 \cdot 3^{25} \approx 1.7 \cdot 10^{12}$ configurations – 2 possibilities for the active player, and 3 options for each cell in the grid. This number is in a similar order of magnitude as the numbers of positions in Connect-Four, which was solved 30 years ago [1].

However, the game "tree" of Quixo is very different from other similar game "trees." For most games, the "trees" are directed acyclic graphs (DAG), assuming the merging of identical positions reached from different histories. Conversely, for Quixo, the game "tree" contains cycles. Indeed, especially in the later stages of the game, when the grid is mostly full of Xs and Os, most moves do not add symbols, only reorganize them. Therefore, a simple minimax algorithm (with or without alpha-beta pruning) may never terminate.

As a consequence, instead of searching the game "tree" with a DFS algorithm (as done by minimax or alike algorithms), it is necessary to use a more costly approach. Ideally, one would like to analyze the whole game "tree," but it is currently impossible to store it all at once in memory on commodity hardware.

1.3 Results Overview

Our solution involves a combination of backward-induction and value-iteration algorithms, implemented using a state representation that is both time and space efficient. Based on our computations, the regular 5×5 Quixo is a Draw; neither player has a winning strategy if both players play optimally, and the game continues forever. On smaller grids, the first player wins. Interestingly, on the 4×4 grid, it takes at least 21 moves (11 moves from the first player and 10 from the opponent) to win. Since $21 > 16$, it is always necessary to re-use existing tiles.

Outline. Section 2 presents some basic definitions and terminology used in the paper. Sections 3 and 4 describe respectively the data structures and the algorithms used to solve Quixo. Section 5 summarizes our findings and highlights some unexpected observations. Finally, Sect. 6 concludes the paper with a list of open problems.

2 Preliminaries

By convention, the *first player* is player X and the *second player* is player O. The *board* corresponds to the 25 tiles and the *active player* denotes the player playing next. Note that, contrarily to Tic-Tac-Toe or Connect4, the active player cannot be deduced automatically from a given board. Therefore a *state* of the game consists of both a board and an active player. The *initial state* is the empty board (that is, the board with 25 empty tiles) with player X as the active player.

A state is *terminal* if its board contains a line of Xs or Os tiles. The *children* of a given state are all states obtained by a move of the active player. A terminal state has no children since the game is over and there are no valid moves. The *parents* of a state are defined analogously. The set of states and parent-child relations induce the *game graph* of Quixo (referred to as the game "tree" in the prequel). As mentioned earlier, this graph is neither a tree nor an acyclic graph.

Outcomes. Each state has a *state value*, also called *outcome* which can be either *active-player-Win*, *active-player-Loss*, or *Draw*. For brevity, the *active-player* part is omitted, and a Win (resp. Loss and Draw) state denotes a state whose outcome is Win (resp. Loss and Draw). The outcome of a terminal state is trivially defined. For non-terminal states, the outcome is inductively defined: Win if there is at least one Loss child, Loss if all children are win, Draw otherwise.

Symmetries and Swapping. Rotating or mirroring the board does not change the state value. Therefore states can be grouped in equivalence classes. This optimization divides approximately by eight the number of states: four being due to rotations, and two to mirroring. Also, swapping the active player and flipping all Xs and Os to Os and Xs respectively creates a new equivalent state. Figure 2 illustrates these notions.

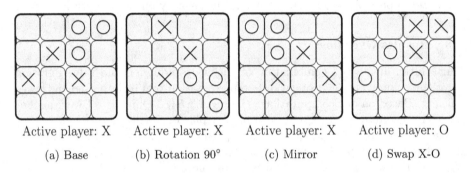

Active player: X   Active player: X   Active player: X   Active player: O

(a) Base   (b) Rotation 90°   (c) Mirror   (d) Swap X-O

Fig. 2. Equivalent states with respect to symmetries and swapping

In the remaining of the paper, all states have X as the active player. By an abuse of notation, we then identify the state and its board, omitting the active player.

3   Solving Quixo – Data Structures

This section considers only 5×5 Quixo but explanations can easily be adapted for smaller grids. First, we describe our memory-efficient representation of states. Then we focus on the more general problem of storing intermediate results. Due to space constraints, some explanations are omitted and can be found in the full paper [7].

3.1   Bitboard State Representation

unused   25 bits to store X locations   unused   25 bits to store O locations

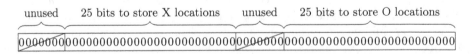

Fig. 3. State representation in 64bits (LSB on the right)

We use 64 bits to encode a state as depicted in Fig. 3. This representation offers some decisive advantages. It enables very fast computation of all basic operations.

s = [0000000]0000001000111010100000100[0000000]0011000100000101000001000

B = 0000000000000111110000000000000000000000000000001111000000000000000

C = 0000000000000000000000000000000000000000000000001000000000000000000

Fig. 4. Our 64bits representation of the state of Fig. 1b and constants B and C used to compute the move creating the state of Fig. 1c using (((s & B) >> 1) & B) | (s & ~B) | C

Given a state s, and some appropriate pre-computed constants A, B, C (see Fig. 4), all the following operations can be executed efficiently (<<, >>, &, |, and ~ denote usual bitwise operations):

- Swapping the players: s << 32 | s >> 32
- Checking the existence of a tile at a given location: s & A != 0
- Checking the existence of a given line of Xs or Os: s & A == A
- More interestingly, moves can also be computed quickly; e.g. for a down-pushing move: (((s & B) >>5) & B) | (s & B) | C

Unfortunately, rotations and symmetries are still costly to compute. In fact, we believe that there is no efficient way to compute rotations with a compact data structure. Based on our observations, it is faster to avoid symmetry optimizations, and simply compute values independently for all symmetrical states. In the next section, we thus investigate how to store the outcomes of 3^{25} states.

3.2 Optimized Storage of Results

Using our optimized state representation, computations can be done quickly. It remains to consider the problem of storing the outcome of each state. Indeed, in order to strongly-solve the game, we need to record the outcome of all possible states. Three possible outcomes (Win/Loss/Draw) means that 2 bits are necessary to store each outcome. Using a typical (state: value) associative array requires at least 64 bits + 2 bits per entry, which sums up to more than 6.5TB.[2]

It is possible to enumerate all possible states in a pre-determined order. It is therefore natural to only store the outcomes in a (giant) bit array. Again, 2 bits per entry yields a total size of $2 \cdot 3^{25} = 197$GB. Although more reasonable, renting a server with 200 GB of RAM may still require a significant investment. We further reduce memory requirements.

The obvious solution is to avoid storing all outcomes at the same time in RAM. Using backward induction (see Sect. 4), we only need to have a subset of already computed values to compute the new outcomes. For example, to compute all states containing either 10 Xs and 8 Os or 8 Xs and 10 Os, it is sufficient to know the (inductively computed) outcomes of states with either 8 Xs and 11 Os,

---

[2] In practice, this amount of space is likely much higher due to memory alignment. So, a more realistic estimation for full storage is in the order of 15 TB.

or 10 Xs and 9 Os. Therefore we partition the 3^{25} states based on the number of Xs and Os. Let $\mathcal{C}_{x,o}$ denotes the class of states containing x Xs and o Os.

The largest class is $\mathcal{C}_{8,8}$ which contains $\binom{25}{8} \cdot \binom{17}{8} \approx 2.6 \cdot 10^{10}$ states. Using 2 bits per state, it corresponds to ≈ 6.1 GB of RAM. Since we can implement our algorithms using at most two classes loaded in memory at once, it becomes possible to solve the game on a more typical 16 GB-RAM computer. While this partitioning is easy, there is an hidden problem:

Creating a bijection between the set of all 3^{25} states and the set of natural numbers $\{0, \ldots, 3^{25} - 1\}$ is straightforward and easily computable (in both directions); one can see the 25 cells as the 25-digit ternary representation of a number (0 for empty, 1 for X, and 2 for O).

However, creating a bijection between $\mathcal{C}_{x,o}$ and the set $\{0, \ldots, \binom{25}{x} \cdot \binom{25-x}{o} - 1\}$ is less straightforward and more difficult to implement in an efficient way. Due to space constraints, implementation details are omitted here.

4   Solving Quixo – Algorithms

4.1   Computing Outcome

Value Iteration. After designing data structures, we now present our algorithms. As explained in Sect. 1.2, due to cycles in the game "tree," minimax algorithm cannot be used for Quixo. The most natural algorithm for solving such games is the *Value Iteration* (VI) algorithm (recalled in Algorithm 1). In the pseudo-code, the children of a state denote the set of states that are reachable after one move.[3] This algorithm follows closely the definition of outcome provided in Sect. 2.

Algorithm 1. Value Iteration (VI)

1: **for all** states s **do**
2:      **if** there is a line of Xs in s **then**
3:          outcome[s] $\leftarrow$ *Win*
4:      **else if** there is a line of Os in s **then**
5:          outcome[s] $\leftarrow$ *Loss*
6:      **else**
7:          outcome[s] $\leftarrow$ *Draw*
8: **repeat**
9:      **for all** states s such that outcome[s] = *Draw* **do**
10:         **if** at least one child of s is Loss **then**
11:             outcome[s] $\leftarrow$ *Win*
12:         **else if** all children of s are Win **then**
13:             outcome[s] $\leftarrow$ *Loss*
14: **until** no update in the last iteration

---
[3] Some implementation details are omitted. For example, since we only consider states with active player X, we need to swap the tiles/players after each move.

Backward Induction. Quixo cannot be solved directly applying this algorithm since it would require to store all outcomes at once in RAM (and thus would be too slow due to memory caching). Fortunately, we can use the classes $\mathcal{C}_{x,o}$ we defined in Sect. 3.2. Indeed, for any state of $\mathcal{C}_{x,o}$, its children belong to $\mathcal{C}_{o,x} \bigcup \mathcal{C}_{o,x+1}$ (due to player swap after each move). Thus, it becomes possible to compute all outcomes of $\mathcal{C}_{x,o} \bigcup \mathcal{C}_{o,x}$ using only $\mathcal{C}_{x,o+1} \bigcup \mathcal{C}_{o,x+1}$. Starting from states with 25 non-empty tiles, and using *backward induction*, we can compute all outcomes having only four classes of states in RAM at any time. The corresponding pseudo-code is given in Algorithm 2.

Algorithm 2. Backward Induction using VI internally

1: **for** $n = 25$ **to** 0 **do**
2:   **for** $x = 0$ **to** $\lceil n/2 \rceil$ **do**
3:     $o \leftarrow n - x$
4:     **if** $n < 25$ **then**
5:       Load outcomes of classes $\mathcal{C}_{x,o+1}$ and $\mathcal{C}_{o,x+1}$
6:       Compute outcomes of classes $\mathcal{C}_{x,o}$ and $\mathcal{C}_{o,x}$ using VI
7:       Save outcomes of classes $\mathcal{C}_{x,o}$ and $\mathcal{C}_{o,x}$
8:       Unload all outcomes

This algorithm is likely to be able to solve Quixo, but, in practice, it is too slow. Unfortunately, it is difficult to evaluate precisely its complexity because it depends on the number of internal (value) iterations, which is itself difficult to predict. The topology of the game "tree" has a strong impact on the required number of iterations to converge.

Algorithmic Optimizations. We propose two algorithmic enhancements that significantly reduce the computation time. Due to space constraints, pseudocodes and detailed explanations are omitted and available in the full paper.

Use Parent Link. To be a Win state, there should be at least one Loss child. Reversing this statement, we obtain that every parent of a Loss state is a Win state. This simple observation can be used to improve the computation. As soon as a state is found to be a Loss, we can compute all its parents and update their outcome to Win. Eventually they would have been updated to Win in Algorithm 1, but updating them immediately makes it possible to skip searching if states are Win (Lines 10 to 11) and may allow other states to be updated faster too. Note that parents of a given state can also be computed efficiently using a similar method as for computing its children (see Sect. 3.1).

Use Win-or-Draw Outcome. Internal iterations require checking the outcomes of all children of a given state (see Lines 10 and 12 of Algorithm 1). Some of the children belong to already inductively-computed classes, while the others belong to classes currently being computed. More explicitly, for a state $s \in \mathcal{C}_{x,o}$, some children belong to $\mathcal{C}_{o,x}$ and some belong to $\mathcal{C}_{o,x+1}$. The outcome of this latter class has been already computed inductively. It is possible to check them only once by introducing a *new temporary outcome*: WinOrDraw.

Complete Algorithm. Combining value iteration, Backward induction, and our two optimizations, we obtain the complete algorithm we used to solve Quixo. Note that we also added some parallelization to reduce computation time (see [7]).

4.2   Deriving an Optimal Strategy

Using previously described algorithms, it is possible to compute all state outcomes. One may think that always choosing deterministically a winning move (if available) is a winning strategy. Unfortunately, such a strategy does not guarantee winning since Quixo game tree contains cycles. Hence, it is possible to enter a cycle where all states outcomes are Win (for one of the player), yet the game never finishes. Note that this issue does not exist in games without cycles, such as Connect-four.

It is possible to devise a probabilistically winning strategy by choosing a winning move uniformly at random (among winning moves): indeed, from any Win state, there exists a sequence of steps that does not belong to an infinite cycle. So, in an expected finite number of steps, the player wins.

However, the random strategy is not optimal with respect to the number of steps taken to win. Instead, we focus on the strategy to win in the minimum number of steps (assuming the loosing player always chooses the action that delays her loss as much as possible). To actually compute the steps to win or lose, we now store the number of steps to the final outcome, using a new *step* variable. The *step* variable is defined as follows:

– In a terminal state, *step* is 0,
– If the state is Win, *step* is one plus the minimum of the *step*s of Loss children,
– If the state is Loss *step* is one plus the maximum of the *step*s of Win children.

Previous algorithms can be adapted to compute this additional *step* variable. Note that it incurs a memory cost; instead of only 2 bits per outcome, 16 or 32 more bits are needed for an integral type. Since 5×5 Quixo is a Draw (Sect. 5), there is obviously no (optimal) winning strategies. Hence we used this modified algorithm only for the smaller variants of Quixo on 3×3 and 4×4 grids.

Algorithm 3 is the update of Algorithm 1 including the computation of steps.

4.3   Implementations

To increase confidence in our results, all computations (except the one described in Sect. 4.2) have been computed and verified with two independent implementations using slightly different optimizations. The complete source code is available in a public Github repository [5].

Algorithm 3. Value Iteration with steps to Win or Loss

```
 1: for all states s do
 2:     if there is a line of Xs in s then
 3:         outcome[s] ← Win
 4:         step[s] ← 0
 5:     else if there is a line of Os in s then
 6:         outcome[s] ← Loss
 7:         step[s] ← 0
 8:     else
 9:         outcome[s] ← Draw
10: i ← 1
11: repeat
12:     for all states s such that outcome[s] = Draw do
13:         if at least one child of s exists whose outcome[c] is Loss and step[c] is i − 1
            then
14:             outcome[s] ← Win
15:             step[s] ← i
16:         else if all children of s are Win then
17:             outcome[s] ← Loss
18:             step[s] ← max(children steps) + 1
19:     i ← i + 1
20: until no update in last iteration
```

5 Results

5.1 5 × 5 Quixo

Our main result is that *Quixo is a Draw game*. In other words, if perfect players play the game, no one wins, that is, the game never finishes.

Using a single-thread computation,[4] it takes approximately 19 500 min (just under two weeks) to obtain this result. Using multithreading, the running time shrinks to around 1 900 min (i.e. ≈ 32 h) using up to 32 threads. For comparison, it takes around 13.5 s and 0.1 s for 4 × 4 and 3 × 3 grids respectively.

Some Additional Observations. Table 1 shows the total number of Win, Loss, and Draw states. As the number of Draw states is smaller than those of Win or Loss states, it may come as a surprise that the initial state is Draw. However, when looking at the distribution of these states, it appears that most of the Draw states are located near the top of the game "tree" (*i.e.*, with few marked tiles).

Figure 5 displays a few selected states whose outcomes are not trivial.

5.2 4 × 4 Quixo

Contrarily to the real game, the 4 × 4 variant is a Win for the first player. Intuitively, the smaller board makes it easier to create a line. However,

---

[4] We used a Ubuntu 18.04LTS server equipped with 32GB of RAM and powered by a 16-core Intel Core i9-9960X CPU.

Table 1. Total Win, Loss and Draw states numbers

Win	Loss	Draw
441,815,157,309	279,746,227,956	125,727,224,178

  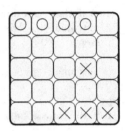

(a) Player X can win. One of the states with the smallest numbers of tiles such that the outcome is not draw.

(b) Player X can win. One of the states with the smallest numbers of tiles such that the outcome is not draw and both players have chosen empty tiles only.

(c) Player X loses. The number of Xs and Os tiles is the same but the active player loses.

Fig. 5. Some interesting states on the 5 × 5 board. Player X is next to play.

winning is not trivial; it requires up to 21 moves when the opponent follows an optimal strategy.

Some additional observations. Some states are obviously not reachable, e.g. a state containing a single O not on an edge. Some other unreachable states are much less obvious, such as the state in Fig. 6a. Globally, there are 41 252 106 reachable states, which accounts for 95.8% of the 3^{16} states. Therefore, ignoring unreachable states in the computation would not be significant.

 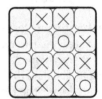

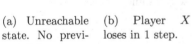

(a) Unreachable state. No previous state.

(b) Player X loses in 1 step.

(c) Player X loses in 22 steps.

(d) Draw state. O can come back to this state (or a symmetric one) with the next O step even if X plays optimally.

Fig. 6. Some interesting states on the 4 × 4 board. Player X is next to play.

Using the algorithm described in Sect. 4.2, we computed the optimal strategies and the numbers of steps required to win/lose.

Typically, a winner wins in an odd number of steps, and a loser loses in an even number of steps. However, some states yield an optimal player to lose in 1 step. One such example is shown in Fig. 6b. In all children states, there is a line of O, so X loses in 1 step.

Another interesting result is that there are some states that lose in 22 steps although no state wins in 23 steps, and the initial state wins in 21 steps. Figure 6c is the only state (and its symmetric states) to lose in 22 steps.

6   Conclusions and Open Questions

To summarize, the official 5×5 Quixo is a Draw game; neither player can win. Smaller 3×3 and 4×4 variants are First-Player-Win games.[5] Given that the 5×5 board is already a Draw game, one may expect larger instances to be Draw games too. We conjecture that it is the case, but we were not able to prove it.

Mishiba and Takenaga proved that a generalization of Quixo is EXPTIME-complete [2]. They consider arbitrary large boards, but players still have to align only five identical symbols. Based on this generalization, a natural question arises; can the first player (or unlikely the second player) create a line of four symbols when playing on the 5×5 board? Changing the two lines losing rule into a winning rule may also change the global outcome.

Finally, a last research direction would be to compute human-playable optimal strategies. We strongly solved Quixo on 4×4 and 5×5 grids. However, playing an optimal strategy remains difficult for humans.

References

1. Allis, L.V.: A knowledge-based approach of connect-four. ICGA J. **11**(4), 165 (1988)
2. Mishiba, S., Takenaga, Y.: 一般化QUIXOの計算複雑さ IEICE general conference p. 26 (2017), (in Japanese)
3. Quixo page on BoardGameGeek website. https://boardgamegeek.com/boardgame/3190/quixo, (accessed 17 September 2021)
4. Quixo page on Gigamic website. https://en.gigamic.com/game/quixo, (accessed: September 17 2021)
5. Quixo source code repository. https://github.com/st34-satoshi/quixo-cpp
6. Tanaka, S., Bonnet, F., Tixeuil, S., Tamura, Y.: Quixo の強解決. In: Proceedings of the 26th Game Programming Workshop, pp. 181–188 (2020), (in Japanese)
7. Tanaka, S., Bonnet, F., Tixeuil, S., Tamura, Y.: Quixo is solved (2020). https://arxiv.org/abs/2007.15895

---

[5] The 3×3 version is not discussed here and left to the reader (Win in 7 moves).

Solving Bicoloring-Graph Games on Rectangular Boards – Part 1: Partisan Col and Snort

Jos W. H. M. Uiterwijk[✉]

Department of Data Science and Knowledge Engineering (DKE),
Maastricht University, Maastricht, The Netherlands
uiterwijk@maastrichtuniversity.nl

Abstract. In this paper we give an overview of results obtained for solving the combinatorial games of Col and Snort on rectangular boards.

For Col on boards with at least one dimension even we give a strategy guaranteeing a win for the second player. For Col on general odd × odd boards we found no applicable strategy, though all experimental data show second-player wins. For Linear Col we were able to prove using Combinatorial Game Theory (CGT) that all chains, including odd-length chains, are second-player wins.

A similar strategy as for Col guarantees for Snort on boards with both dimensions even a win for the second player and with at least one dimension odd a first-player win. Snort therefore is completely solved.

1 Introduction

In Artificial Intelligence map-coloring has been a prime focus of research. In its basic form the question is: can a map with neighboring regions be colored with some finite number of colors, such that neighboring regions are colored differently? Any map-coloring problem is equivalent with some graph-coloring problem, where nodes represent regions, and edges denote common frontiers between corresponding regions, and the goal is to color all nodes in the graph such that any two connected nodes are colored differently.

In the field of Combinatorial Game Theory (CGT in short), graph-coloring problems can be transformed into games by changing the goal of a game: not to fully color a map, but to make the last move (under the normal ending rule) when players alternately color one region. It is common to restrict such graph-coloring games to two colors, where both players have their own color, conventionally Black for the player who starts the game and White for the opponent.

The two combinatorial graph-coloring games most well known are surely Col and Snort, both first analyzed by Conway [3]. He attributed Col to Colin Vout and Snort to Simon Norton. Both are similar in the sense that both players alternately color a node in the graph, where one player may only color it black, the other only white. The two games differ in their conditions for coloring: in Col neighboring nodes may not be colored the same (further called the Col-condition), while in Snort they may not be colored differently (the Snort-condition).

© Springer Nature Switzerland AG 2022
C. Browne et al. (Eds.): ACG 2021, LNCS 13262, pp. 96–106, 2022.
https://doi.org/10.1007/978-3-031-11488-5_9

Although both games can be played on any types of graphs, in this paper we concentrate on rectangular boards (sometimes just refered to as *boards*), where both players alternately put a stone of their color on a square. As a special case of Col and Snort on boards we consider Linear Col and Snort, played on one-dimensional boards (further called *chains*).

The literature on Col and Snort is very scarce. It has been introduced in the framework of CGT in the seminal books *On Numbers and Games* [3] and *Winning Ways* [2], where many small graphs are given and some more general rules are exemplified. Most of these are irrelevant for analyzing larger boards, except Linear Col, for which values were given without proof. Such a proof is given in this paper. Recently, a bachelor thesis by Demeur [4] reports solving many Col and Snort boards with sizes up to some 30 squares, based on $\alpha\beta$ search. We are not aware of any further analyses of Col and Snort.

2   Combinatorial Game Theory for Col and Snort

In this section we give a short introduction to the Combinatorial Game Theory as far as relevant for Col and Snort. For a more thorough introduction, we refer to the literature, in particular [1–3,6].

In a combinatorial game, the players are conventionally called Left and Right. For Col and Snort, Left is the player moving the black stones, therefore also denoted as Black, and similarly Right (White) moves the white stones. A game (position) G is then represented as $G = \{G^L \mid G^R\}$, where G^L and G^R stand for sets of games (the *options*) that players Left and Right, respectively, can reach by making one move in the game. The *value* of a game indicates how good a game is for a player. Then there are four possible outcome classes.

1. The class $\mathcal{L}$ consists of all positions where Left wins, irrespective of who moves first. These positions have strictly positive values.
2. The class $\mathcal{R}$ consists of all positions where Right wins, irrespective of who moves first. These positions have strictly negative values.
3. The class $\mathcal{N}$ consists of all positions where the player to move (the next player) wins. These positions have fuzzy values (incomparable with 0).
4. The class $\mathcal{P}$ consists of all positions where the player to move loses, so the previous player wins. These positions all have value 0.

Depending on the outcome class of a game, several types of values are possible. We treat the most important ones for Col and Snort in the next subsections.

2.1   Numbers and Star

Numbers have the property that any option is a number itself, and that any left option has a lower value than any right option. The simplest number game is the endgame $\{\mid\}$, denoted as 0. In this position, no player has any available moves, so it is a loss for the player to move. Larger or smaller numbers are built recursively. So $0 = \{\mid\}; 1 = \{0\mid\}; 2 = \{1\mid\}; -1 = \{\mid 0\}; -2 = \{\mid -1\};$ etc.

Some example Col positions with integer values are given in Fig. 1. In the left position, there is only one empty square, but due to the Col-condition it can be colored neither black nor white, so this position has value 0. In the middle position there is also one empty square, which may only be colored black, so this position has value +1. In the right position, with two empty squares, only White can move (twice), so this position has value −2.

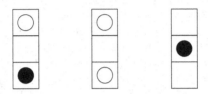

Fig. 1. Example Col positions on the 3 × 1 board with integer values.

Also fractions are possible. For example, the position in Fig. 2 has value $\{-1, 0 | 1\} = \{0 | 1\}$. Naturally this value is notated as $1/2$ (supported by the proof that two games with value $1/2$ are equivalent to one game with value 1).

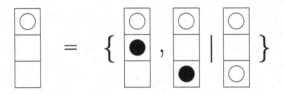

Fig. 2. Example Col position on the 3 × 1 board with value $1/2$.

Besides the endgame 0 on which all numbers are built, the most important simple game is the one denoted as Star or $*$. It is defined as $* = \{0 | 0\}$, where the player to move has just 1 option, leading to the endgame. Therefore, whereas the game 0 is a game where the second player to move wins (since trivially the next player cannot move), the $*$ is a game where the first player to move wins. A trivial example in both Col and Snort is a lone empty square; a more complex example in both games is given in Fig. 3.

Fig. 3. An example $*$ position in both Col and Snort on the 3 × 3 board.

$*$ is a fuzzy number, incomparable with 0. In fact it is a nimber, which is formally defined as $*n = \{*0, *1, *2, \cdots, *(n-1) \mid *0, *1, *2, \cdots, *(n-1)\}$.

In partisan games like Col and Snort nimbers can occur, but are quite rare; so far we only found nimbers $0 = *0$ and $* = *1$. Conway [3] has proven that in Col every position has as value a number (z) or a number plus $*$ (notation $z*$).

2.2   Switches

Simple Snort games often have numbers as options, but have at least one left option with a larger value than some right option. For simple switches of the form $\{a|b\}$ ($a > b$) an alternative notation is $\frac{a+b}{2} \pm \frac{a-b}{2}$, where the first term is the *mean* value of the switch and the second term its *temperature*. A few example Snort positions with simple switches as values are given in Fig. 4.

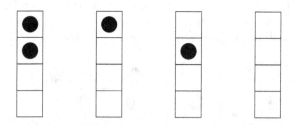

Fig. 4. Example Snort positions on the 4×1 board with switch values.

In the left position Black to move can take any of the remaining two empty squares to reach a position of value $+1$, whereas White's only option is to take the lowest empty square, ending the game; so this position is a switch with value $\{1|0\}$ (alternative notation $1/2 \pm 1/2$). The second and third positions likewise have values $\{2|-1\}$ (alternatively $1/2 \pm 1 1/2$) and $\{2|1\}$ (alternatively $1 1/2 \pm 1/2$). Clearly the third position is to be preferred for Black over the second one. As a consequence, in the rightmost position this option for Black dominates; since White options are the negation of Black's options, the latter position can be seen to have value $\{\{2|1\}|\{-1|-2\}\}$ (alternatively $\pm\{2|1\}$). It is clear that larger boards can have quite long and complicated switches as values.

Note that switches of the form $\pm x$ are called *fair switches*, since both players to move gain the same profit. Obviously, all switches for empty Snort boards are fair switches.

3   Partisan Col on Rectangular Boards

In this section, we investigate standard (*partisan*) Col on rectangular boards. This means that both players have their own stones, black for the Left player and white for the Right player.

3.1   Col on $m \times n$ Boards with m and/or n Even

For $m \times n$ Col boards with m and/or n even we found that the second player always can win. This can easily be proven as stated in Theorem 1.

Theorem 1. *All empty $m \times n$ Col boards with m and/or n even are second-player wins and thus have CGT value 0.*

Proof. The second player can use a copy-strategy as follows. Wherever the first player (Black) moves, the second player (White) plays symmetric wrt the center. Then after every black move the board has opposite-color symmetry wrt the center. Since every black move must fulfil the Col-condition, every white move will automatically fulfil the Col-condition also. Consequently, the second player makes the last move and wins.   □

We further denote this strategy as the *center strategy*. Example Col games on the 4×4 and 4×5 boards where White uses this winning strategy are shown in Fig. 5. The left diagram shows a Col game on an even × even board, the right diagram on an even × odd board. The numbers inside the stones are the move numbers. The small dot indicates the center of the board.

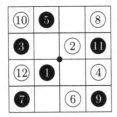 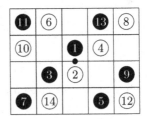

Fig. 5. Example Col games on the 4×4 and 4×5 boards won by White.

3.2   Col on $m \times n$ Boards with m and n Odd, with $m, n \geq 3$

As a consequence of Theorem 1, only $m \times n$ Col boards with both m and n odd are of interest for solving. Obviously, the second player cannot use the center strategy, since the first player can at some moment color the center, to which the second player cannot respond using this strategy. In principle such boards can therefore be either first-player or second-player wins. Only for empty $1 \times n$ boards (see Sect. 3.3) with odd n we know the solutions, which limits the interesting boards to be solved to empty $m \times n$ boards with both m and n odd and $m, n \geq 3$. Demeur [4] proved that the 3×3, 3×5, 3×7, 3×9, and 5×5 boards are second-player wins (CGT value 0), but his analyses show no general applicable winning strategy for either player on odd × odd boards.

3.3   Linear Col

We denote Col on a $1 \times n$ (or equivalently $n \times 1$) board as Linear Col. Of course we already know that Linear Col on even-length boards is a second-player win, so has CGT value 0. The odd-length Linear Col boards are still of interest, since our previous analysis gives no clue. Linear Col is supposedly completely solved, see *Winning Ways* [2], Vol. 1, pp. 49–50. Since there are no proofs given we provide such a proof in Theorem 2. This will be a proof by induction on the length of the chain. To do this we do not only consider just empty, black, or white nodes, but also so-called *tinted* nodes. After coloring a node the Col-condition imposes that its neighbors never may receive the same color. To indicate this we may tint an empty neighbor of a black-colored node white, to indicate that such a node may only be colored White in the future. Likewise, an empty neighbor of a white-colored node is tinted black. If an empty node receives both a black and a white tint, it means that this node may not be colored anymore at all.

We use the following notation for this: **B** for a black-colored square, **W** for a white-colored square, **b** for a black-tinted square, **w** for a white-tinted square, **x** for an empty square that can no longer be colored by either player, and **o** for an empty square that still can be colored by either player. For brevity we omit all edges.

Theorem 2. *Empty Linear Col chains of length n have CGT value 0 for n > 1.*

Proof. Note that as soon as a node is white-colored or black-colored (and the neighbors have been updated), it may be removed from the chain, which accordingly splits. It splits also when an empty node can not be colored by any player, since this node may also be removed. Therefore the CGT value of a Linear Col chain can be determined by the values of shorter subchains, in which only end nodes are possibly tinted. For the chains $\mathbf{o}\cdots\mathbf{o}$, $\mathbf{b}\cdots\mathbf{b}$, and $\mathbf{w}\cdots\mathbf{w}$ we consider only options in the left half of the chain, for other chains we consider all options. Of course symmetric chains like $\mathbf{b}\cdots\mathbf{o}$, and $\mathbf{o}\cdots\mathbf{b}$ have the same values, whereas swapping **b**s and **w**s in a chain yields the negation of the CGT value. **0** denotes the Linear Col chain of zero length, of course having value 0.

The analyses below always proceed in (at most) five steps: 1) determine the options of the chain; 2) simplify the options by removing colored and uncolorable nodes; 3) replace the options by their CGT values; 4) remove dominated options; and 5) determine the CGT value of the original chain.

Base cases: $1 \times n$ chains with $n \le 4$ have the following values:

$n = 1$: $\mathbf{o} = \{\mathbf{B}|\mathbf{W}\} = \{0|0\} = \{0|0\} = *$; $\mathbf{b} = \{\mathbf{B}|\} = \{0|\} = \{0|\} = 1$; similarly $\mathbf{w} = -1$.

$n = 2$: $\mathbf{oo} = \{\mathbf{Bw}|\mathbf{Wb}\} = \{\mathbf{w}|\mathbf{b}\} = \{-1|1\} = 0$; $\mathbf{bo} = \{\mathbf{Bw}, \mathbf{xB}|\mathbf{bW}\} = \{\mathbf{w}, \mathbf{0}|\mathbf{b}\} = \{-1, 0|1\} = \{0|1\} = 1/2$; similarly $\mathbf{ob} = 1/2$, $\mathbf{wo} = \mathbf{ow} = -1/2$; $\mathbf{bb} = \{\mathbf{Bx}|\} = \{0|\} = \{0|\} = 1$; similarly $\mathbf{ww} = -1$; $\mathbf{bw} = \{\mathbf{Bw}|\mathbf{bW}\} = \{\mathbf{w}|\mathbf{b}\} = \{-1|1\} = 0$; similarly $\mathbf{wb} = 0$.

$n = 3$: $\mathbf{ooo} = \{\mathbf{Bwo}, \mathbf{wBw}|\mathbf{Wbo}, \mathbf{bWb}\} = \{\mathbf{wo}, \mathbf{w}+\mathbf{w}|\mathbf{bo}, \mathbf{b}+\mathbf{b}\} = \{-1/2, -2|1/2, 2\} = \{-1/2|1/2\} = 0$; $\mathbf{boo} = \{\mathbf{Bwo}, \mathbf{xBw}, \mathbf{bwB}|\mathbf{bWb}, \mathbf{bbW}\} = \{\mathbf{wo}, \mathbf{w}, \mathbf{bw}|\mathbf{b}+\mathbf{b}, \mathbf{bb}\} = \{-1/2, -1, 0|2, 1\} = \{0|1\} = 1/2$; similarly $\mathbf{oob} = 1/2$, $\mathbf{woo} = \mathbf{oow} = -1/2$; $\mathbf{bob} = \{\mathbf{Bwb}, \mathbf{xBx}|\mathbf{bWb}\} = \{\mathbf{wb}, \mathbf{0}|\mathbf{b}+\mathbf{b}\} = \{0, 0|2\} = \{0|2\} = 1$; similarly $\mathbf{wow} = -1$; $\mathbf{bow} = \{\mathbf{Bww}, \mathbf{xBw}|\mathbf{bWx}, \mathbf{bbW}\} = \{\mathbf{ww}, \mathbf{w}|\mathbf{b}, \mathbf{bb}\} = \{-1, -1|1, 1\} = \{-1|1\} = 0$; similarly $\mathbf{wob} = 0$.

$n = 4$: $\mathbf{oooo} = \{\mathbf{Bwoo, wBwo|Wboo, bWbo}\} = \{\mathbf{woo, w + wo|boo, b + bo}\} = \{-1/2, -1 1/2|1/2, 1 1/2\} = \{-1/2|1/2\} = 0$; $\mathbf{booo} = \{\mathbf{Bwoo, xBwo, bwBw, bowB| bWbo, bbWb, bobW}\} = \{\mathbf{woo, wo, bw + w, bow|b + bo, bb + b, bob}\} = \{-1/2, -1/2, -1, 0|1 1/2, 2, 1\} = \{0|1\} = 1/2$; similarly $\mathbf{ooob} = 1/2$, $\mathbf{wooo} = \mathbf{ooow} = -1/2$; $\mathbf{boob} = \{\mathbf{Bwob, xBwb|bWbb}\} = \{\mathbf{wob, wb|b + bb}\} = \{0, 0|2\} = \{0|2\} = 1$; similarly $\mathbf{woow} = -1$; $\mathbf{boow} = \{\mathbf{Bwow, xBww, bwBw|bWbw, bbWx, bobW}\} = \{\mathbf{wow, ww, bw + w|b + bw, bb, bob}\} = \{-1, -1, -1|1, 1, 1\} = \{-1|1\} = 0$; similarly $\mathbf{woob} = 0$.

So for $1 \leq n \leq 4$ we have

$$\mathbf{o} = *, \mathbf{b} = 1, \mathbf{w} = -1, \mathbf{o} \cdots \mathbf{o} = 0,$$
$$\mathbf{b} \cdots \mathbf{o} = \mathbf{o} \cdots \mathbf{b} = 1/2, \mathbf{w} \cdots \mathbf{o} = \mathbf{o} \cdots \mathbf{w} = -1/2, \tag{1}$$
$$\mathbf{b} \cdots \mathbf{b} = 1, \mathbf{w} \cdots \mathbf{w} = -1, \mathbf{b} \cdots \mathbf{w} = \mathbf{w} \cdots \mathbf{b} = 0$$

Induction hypothesis: suppose Eq. (1) holds for chains of length up to $k - 1$. **Induction steps:** consider a chain of length $k \geq 5$. We then have the following subcases, where a '$\cdots$' now indicates a sequence of nodes $\mathbf{o}$, not of arbitrary length, but the length needed to have a complete chain of length k. For entries with chains '$\cdots$' at both sides of the colored square a range of possible entries is meant such that all combinations of left and right lengths are included with always a total length of k.

$\mathbf{o} \cdots \mathbf{o}$: B moves gives $\{\mathbf{Bw} \cdots \mathbf{o}, \mathbf{wBw} \cdots \mathbf{o}, \mathbf{o} \cdots \mathbf{wBw} \cdots \mathbf{o}\} = \{\mathbf{w} \cdots \mathbf{o}, \mathbf{w} + \mathbf{w} \cdots \mathbf{o}, \mathbf{o} \cdots \mathbf{w} + \mathbf{w} \cdots \mathbf{o}\} = \{-1/2, -1 1/2, -1\} = \{-1/2\}$. Similarly, W moves gives $\{1/2\}$. So $\mathbf{o} \cdots \mathbf{o} = \{-1/2|1/2\} = 0$.

$\mathbf{b} \cdots \mathbf{o}$: B moves gives $\{\mathbf{Bw} \cdots \mathbf{o}, \mathbf{xBw} \cdots \mathbf{o}, \mathbf{b} \cdots \mathbf{wBw} \cdots \mathbf{o}, \mathbf{b} \cdots \mathbf{wBw}, \mathbf{b} \cdots \mathbf{wB}\} = \{\mathbf{w} \cdots \mathbf{o}, \mathbf{w} \cdots \mathbf{o}, \mathbf{b} \cdots \mathbf{w} + \mathbf{w} \cdots \mathbf{o}, \mathbf{b} \cdots \mathbf{w} + \mathbf{w}, \mathbf{b} \cdots \mathbf{w}\} = \{-1/2, -1/2, -1/2, -1, 0\} = \{0\}$. W moves gives $\{\mathbf{bWb} \cdots \mathbf{o}, \mathbf{b} \cdots \mathbf{bWb} \cdots \mathbf{o}, \mathbf{b} \cdots \mathbf{bWb}, \mathbf{b} \cdots \mathbf{bW}\} = \{\mathbf{b} + \mathbf{b} \cdots \mathbf{o}, \mathbf{b} \cdots \mathbf{b} + \mathbf{b} \cdots \mathbf{o}, \mathbf{b} \cdots \mathbf{b} + \mathbf{b}, \mathbf{b} \cdots \mathbf{b}\} = \{1 1/2, 1 1/2, 2, 1\} = \{1\}$. So $\mathbf{b} \cdots \mathbf{o} = \{0|1\} = 1/2$. Similarly $\mathbf{o} \cdots \mathbf{b} = 1/2$, $\mathbf{w} \cdots \mathbf{o} = \mathbf{o} \cdots \mathbf{w} = -1/2$.

$\mathbf{b} \cdots \mathbf{b}$: B moves gives $\{\mathbf{Bw} \cdots \mathbf{b}, \mathbf{xBw} \cdots \mathbf{b}, \mathbf{b} \cdots \mathbf{wBw} \cdots \mathbf{b}\} = \{\mathbf{w} \cdots \mathbf{b}, \mathbf{w} \cdots \mathbf{b}, \mathbf{b} \cdots \mathbf{w} + \mathbf{w} \cdots \mathbf{b}\} = \{0, 0, 0\} = \{0\}$. W moves gives $\{\mathbf{bWb} \cdots \mathbf{b}, \mathbf{b} \cdots \mathbf{bWb} \cdots \mathbf{b}\} = \{\mathbf{b} + \mathbf{b} \cdots \mathbf{b}, \mathbf{b} \cdots \mathbf{b} + \mathbf{b} \cdots \mathbf{b}\} = \{2, 2\} = \{2\}$. So $\mathbf{b} \cdots \mathbf{b} = \{0|2\} = 1$. Similarly $\mathbf{w} \cdots \mathbf{w} = -1$.

$\mathbf{b} \cdots \mathbf{w}$: B moves gives $\{\mathbf{Bw} \cdots \mathbf{w}, \mathbf{xBw} \cdots \mathbf{w}, \mathbf{b} \cdots \mathbf{wBw} \cdots \mathbf{w}, \mathbf{b} \cdots \mathbf{wBw}\} = \{\mathbf{w} \cdots \mathbf{w}, \mathbf{w} \cdots \mathbf{w}, \mathbf{b} \cdots \mathbf{w} + \mathbf{w} \cdots \mathbf{w}, \mathbf{b} \cdots \mathbf{w} + \mathbf{w}\} = \{-1, -1, -1, -1\} = \{-1\}$. Similarly W moves gives $\{1\}$. So $\mathbf{b} \cdots \mathbf{w} = \{-1|1\} = 0$. Similarly $\mathbf{w} \cdots \mathbf{b} = 0$.

This means that based on the assumption that Eq. (1) holds for chain length $k - 1$ it follows that it holds for chain length k. Combined with the base cases, Eq. (1) consequently holds for arbitrary length chains. □

Concludingly, all empty $1 \times n$ Col boards are second-player wins (CGT value 0), except the 1×1 board has value $*$, and so is a trivial first-player win.

4 Partisan Snort on Rectangular Boards

Although standard (partisan) Col and Snort are very similar games, it turns out that they differ considerably in CGT outcomes and values. In this section we focus on Snort, again on rectangular boards.

4.1 Snort on $m \times n$ Boards with m and n Even

For $m \times n$ Snort boards with m and n both even we found that the second player always can win. This can easily be proven as stated in Theorem 3.

Theorem 3. *All empty $m \times n$ Snort boards with m and n even are second-player wins and thus have CGT value 0.*

Proof. White as second player follows the center strategy. So after every black move, White maintains opposite colored squares wrt the center of the board, meaning that White necessarily makes the last move and wins.   □

Note that this strategy is exactly the same as used in Col on boards with at least one dimension even. Although the Snort-condition differs, for even × even boards the symmetry applied makes sure that after any black move obeying the Snort-condition the white response automatically also obeys this condition.

An example Snort game on the 4 × 4 board is shown in Fig. 6. The first eight moves are the same as the Col game shown in Fig. 5. This is possible since for these moves it holds that there are no colored neighbors yet. From the ninth move on every move necessarily has a colored neighbor and therefore the Snort game now differs from the Col game.

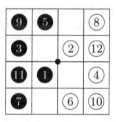

Fig. 6. Example Snort game on the 4 × 4 board won by White.

4.2 Snort on $m \times n$ Boards with m and/or n Odd

For $m \times n$ Snort boards with m and/or n odd the second player cannot use the above copy-strategy to win the game. Instead, we found that the first player always can win, see Theorem 4.

Theorem 4. *All empty $m \times n$ Snort boards with m and/or n odd are first-player wins.*

Proof. First assume that both m and n are odd. Black as first player then starts coloring the single center square and then follows the center strategy. Although the black center inhibits its neighbors to be colored white in the future, it does not hamper Black, so the center strategy still is always possible. Therefore, Black can maintain opposite colored squares wrt the center after every white move (of course excluding the center). Therefore Black makes the last move and wins.

When only one of m and n is odd (arbitrarily suppose m), then the number of rows is odd and the center of the board is in the middle row between the two middle squares. Now Black as first player colors one of these two middle squares and then again can use the center strategy. Of course White cannot use the second middle square. So the center strategy again guarantees Black to make the last move and win.   □

The latter result, stating that the first player wins on a board with at least one dimension odd, does not give the CGT value of these Snort boards, which in principle can be any fuzzy value (like a fair switch or a nimber). Example games where the first player uses this winning strategy are given in Fig. 7.

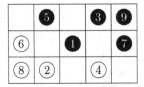 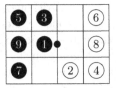

Fig. 7. Example Snort games on the 3×5 and 3×4 boards won by Black.

As a consequence of Theorems 3 and 4 strategically solving rectangular Snort boards is of no more interest, since the dimensions of the board fully determine the winner. Regarding full (CGT) values, determining values of rectangular Snort boards with at least one dimension odd is still of interest.

4.3   Linear Snort

Since $1 \times n$ chains are instances of odd × even or odd × odd boards, and since both these board categories are first-player wins for Snort, we know that all Linear Snort boards are first-player wins with fuzzy values, like nimbers or fair switches. To see if we can find some pattern we determined many CGT values for Linear Snort, using the CGSUITE system [7]. The following values where obtained for various lengths n of the board: $n = 1$: $*$; $n = 2$: ± 1; $n = 3$: ± 2; $n = 4$: $\pm\{2|1\}$; $n = 5$: $\pm(1,\{3|0\})$; $n = 6$: $*$; $n = 7$: $\pm(1,\{4|3||*|-1*\},\{4|3||\pm 1,\{1*|*\}\})$; $n = 8$: $\pm\{\{5|2\},\{5|2*\}|\pm 2,\{2|1||0|-1\},\{2*|-2\}\}$; and $n = 9$: $\pm(2*)$. For lengths 10 to 12 we found fair switches with canonical forms consisting of 273, 628, and 1954 symbols respectively, which we do not reproduce here. Unsurprisely we did not find any pattern in these CGT values.

5   Conclusions and Future Research

We summarize our main results in Table 1. In this table, for every board type we give the outcome class.'?' indicates that the outcome in general is unknown. Instances of outcome class $\mathcal{P}$ have CGT value 0, while instances of outcome class $\mathcal{N}$ have fuzzy CGT values (nimbers or fair switches). For Linear Col and Snort we summarize the results in Table 2.

For odd × odd Col we do know that some instances are second-player wins, but do not know if first-player wins also occur.

Table 1. Outcome classes for Col and Snort on boards of different types.

Game	even × even	odd × even	odd × odd
Col	$\mathcal{P}$	$\mathcal{P}$	?
Snort	$\mathcal{P}$	$\mathcal{N}$	$\mathcal{N}$

Table 2. Outcome classes for Linear Col and Snort on chains of length $n > 1$.

Game	even n	odd n
Linear Col	$\mathcal{P}$	$\mathcal{P}$
Linear Snort	$\mathcal{N}$	$\mathcal{N}$

All results in [2–4] fully support our results. Also, all values in this paper were checked with the CGSUITE system [7], and no discrepancies were found.[1]

For future research we will focus on finding optimal strategies for odd × odd Col boards with both dimensions ≥ 3. We are also interested in results for Col and Snort played on other graphs than rectangular boards. Moreover, we are interested in other bicoloring games. For impartial versions of Col and Snort (dubbed iCol and iSnort) played on rectangular boards we already performed such an analysis [8].

Acknowledgement. We greatfully acknowledge our cooperation with Ewan Demeur. Some of the strategies found were inspired by and formulated during discussions of his thesis research and lead to a deeper insight into the topics of this paper.

References

1. Albert, M.H., Nowakowski, R.J., Wolfe, D.: Lessons in Play: An Introduction to Combinatorial Game Theory. A K Peters, Wellesley (2007)

---

[1] We greatfully used the CGSUITE code for Snort on a rectangular board, as provided in Svenja Huntemann's Ph.D. thesis [5]. The CGSUITE code for the board implementation of Col was a simple adaptation thereof.

2. Berlekamp, E.R., Conway, J.H., Guy, R.K.: Winning Ways for your Mathematical Plays. Academic Press, London (1982). 2nd edition, in four volumes: vol. 1 (2001), vols. 2, 3 (2003), vol. 4. A K Peters, Wellesley (2004)
3. Conway, J.H.: On Numbers and Games. Academic Press, London (1976)
4. Demeur, E.: Solving COL and SNORT on different graphs efficiently. Bachelor thesis, Department of Data Science and Knowledge Engineering, Maastricht University, Maastricht, The Netherlands (2020)
5. Huntemann, S.: The class of strong placement games: complexes, values, and temperature. Ph.D. thesis, Dalhousie University, Halifax, Canada (2018)
6. Siegel, A.N.: Combinatorial Game Theory, Vol. 146 in Graduate Studies in Mathematics. American Mathematical Society (2013)
7. Siegel, A.N.: Combinatorial game suite: a computer algebra system for research in combinatorial game theory (2020). http://cgsuite.sourceforge.net/
8. Uiterwijk, J.W.H.M.: Solving bicoloring-graph games on rectangular boards - part 2: impartial Col and Snort. In: Browne, C., et al. (eds.) ACG 2021. LNCS, vol. 13262, pp. 107–117. Springer, Cham (2022)

Solving Bicoloring-Graph Games on Rectangular Boards – Part 2: Impartial Col and Snort

Jos W. H. M. Uiterwijk$^{(\boxtimes)}$

Department of Data Science and Knowledge Engineering (DKE),
Maastricht University, Maastricht, The Netherlands
uiterwijk@maastrichtuniversity.nl

Abstract. As a sequel to an investigation of the standard (*partisan*) versions of Col and Snort on rectangular boards, we defined *impartial* versions of both games (dubbed *iCol* and *iSnort*). These have the same coloring conditions as their partisan versions, but either player is allowed to use at any move a black or a white stone. For these two games similar strategies show that with both dimensions odd the first player can win, otherwise it is a second-player win. For both Linear versions, analyses using Combinatorial Game Theory show that the even-length chains have value 0, the odd-length chains value $*$.

1 Introduction

In a previous paper [8] we analyzed two well-known bicoloring-graph games on rectangular boards, namely Col and Snort. They were the standard (partisan) versions of these combinatorial games. Both are similar in the sense that both players alternately color a node in the graph, where one player may only color it black, the other only white. The two games differ in their conditions for coloring: in Col neighboring nodes may not be colored the same (the Col-condition), while in Snort they may not be colored differently (the Snort-condition). These two games then were largely solved.

In the present paper we introduce impartial versions of both games, denoted as iCol and iSnort, again focussing on rectangular boards. The games are played with the same restrictions on coloring neighboring nodes as their partisan versions (the Col- and Snort-conditions), but differ in the property that both players always may use either color. This on one hand makes playing them easier, since values of games belong to just two outcome classes (see Sect. 2), but on the other hand makes them more complex, since for neither player it is possible to build significant advantages due to the nature of the games.

Since these games are new to our knowledge, there is no previous scientific literature on them. We only found a single mention of iCol, under the name Bichrome [7], though it was just presented as a fun game and not analyzed mathematically, notably not in the framework of the Combinatorial Game Theory. For iSnort we found no mention in the literature at all.

© Springer Nature Switzerland AG 2022
C. Browne et al. (Eds.): ACG 2021, LNCS 13262, pp. 107–117, 2022.
https://doi.org/10.1007/978-3-031-11488-5_10

2   Combinatorial Game Theory for iCol and iSnort

In this section we give a short introduction to the Combinatorial Game Theory (CGT in short) as far as relevant for the games discussed in this paper. For a more thorough introduction, we refer to the literature, in particular [1,2,4,5].

In a combinatorial game, the players are conventionally called Left and Right. Left starts the game. A game (position) G is then represented by its left and right *options* G^L and G^R, so $G = \{G^L \mid G^R\}$. In this representation, G^L and G^R stand for sets of games that players Left and Right, respectively, can reach by making one move in the game. The *value* of a game indicates how good a game is for a player, where positive values indicate an advantage for Left and negative values an advantage for Right. Then there are four possible outcome classes.

1. The class $\mathcal{L}$ consists of all positions where Left wins, irrespective of who moves first. These positions have strictly positive values.
2. The class $\mathcal{R}$ consists of all positions where Right wins, irrespective of who moves first. These positions have strictly negative values.
3. The class $\mathcal{N}$ consists of all positions where the player to move (the next player) wins. These positions have fuzzy values (incomparable with 0).
4. The class $\mathcal{P}$ consists of all positions where the player to move loses, so the previous player wins. These positions all have value 0.

For impartial games, like iCol and iSnort, it holds that they can only take *nimbers* as values and hence that all positions have outcome class $\mathcal{N}$ or $\mathcal{P}$.

2.1   Nimbers

The simplest nimber game is the endgame $\{\mid\}$, denoted as $*0$. In this position, no player has any available moves, so it is a loss for the player to move and hence a second-player win. Its outcome class is therefore $\mathcal{P}$. Note that this game is the only game being both a nimber and a number, hence $*0 = 0$.

Besides the endgame $*0$, the most important simple game is the one denoted as $*1$, often notated as just $*$. It is defined as $* = \{0 \mid 0\}$, where the player to move has just one option, leading to the endgame. Therefore, whereas 0 is a game where the second player to move wins (since trivially the next player cannot move), $*$ is a game where the first player to move wins. A trivial example in both iCol and iSnort is a lone empty square. Nimbers take their name from the values that can occur in the Nim game [3], where each player has the same options. They are formally defined as $*n = \{*0, *1, *2, \cdots, *(n-1) \mid *0, *1, *2, \cdots, *(n-1)\}$. In case that not all options for a player are consecutive nimbers starting from $*0$, it follows from CGT that the Mex() function applied to the options gives the nimber value of the parent game. The Mex() function (*Minimal excludant*) is the lowest non-negative integer **not** in a set of integers. In case of sums of nimbers they are added pairwise using the Nim-addition rule, which effectively boils down to exclusive-oring the binary representations of the nimbers.

All nimbers other than $*0$ are fuzzy (incomparable with 0) and denote first-player wins. Their outcome class is therefore $\mathcal{N}$.

3   Impartial Col and Snort

We noted in our previous research [8] that most Col and Snort games on rect-angular boards (except Col on odd × odd boards) have known outcomes and easy strategies guaranteeing these outcomes. We then were interested to see if such winning strategies are also possible for impartial versions of Col and Snort. These are defined as follows.

Definition 1. *Impartial Col (iCol for short) and impartial Snort (iSnort) are coloring games on graphs, where the same restrictions on possible colors of neigh-boring nodes apply as in Col and Snort, respectively, but both players are free to use any of the two colors (Black or White) on their turn.*

By this definition both players have exactly the same possible moves in any game position, and so are truly impartial games. We therefore further do not use Black and White for the names of the players in iCol and iSnort, but Left (first to move) and Right (second to move). As stated in Sect. 2 all impartial games, including iCol and iSnort, have only nimbers as possible values.

3.1   iCol on Rectangular Boards

For $m \times n$ iCol boards with m and/or n even the second player always can win. This is proven in Theorem 1.

Theorem 1. *All empty $m \times n$ iCol boards with m and/or n even are second-player wins and thus have CGT value 0.*

Proof. The second player can use a center strategy similar as in Col, i.e. the second player always moves symmetric wrt the center of the board using the opposite color as the previous move. Therefore, after every second-player's move the board is center-symmetric with opposite colors. Consequently, the second player makes the last move and wins.   □

Since in iCol (and later iSnort) both players can use both colors, we add the term "same" or "opp" to the strategy name, so the winning strategy described in the above theorem is called the *center-opp strategy*. Of course when the first player just sticks to using one color, we have a standard Col game won by the second player. Example iCol games on the 4 × 4 and 4 × 5 boards where Right uses this winning strategy are shown in Fig. 1.

The left diagram shows an iCol game on an even × even board, the right diagram on an even × odd board. Right has chosen to always use the center-opp strategy, guaranteeing the win. Note that for iCol on an even × even board an alternative winning strategy for the second player would be to use the center-same strategy. For odd × even and even × odd boards this strategy is not possible, since it might violate the Col-condition.

For $m \times n$ iCol boards with m and n odd the first player always can win. This is proven in Theorem 2.

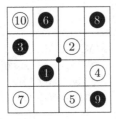 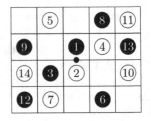

Fig. 1. Example iCol games on the 4 × 4 and 4 × 5 boards won by Right.

Theorem 2. *All empty* $m \times n$ *iCol boards with* m *and* n *odd are first-player wins.*

Proof. Contrary to Col, in iCol the first player can easily win by first coloring the center square arbitrarily (say, black), followed by using the center-same strategy. This guarantees the first player to make the last move and win.   □

An example game where Left uses this strategy to win the game is given in Fig. 2.

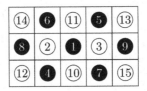

Fig. 2. Example iCol game on the 3 × 5 board won by Left.

Like in Col [8] we only know that odd × odd boards have fuzzy values, but since all values in iCol must be nimbers, we know that the values of odd × odd boards have nimber values $*n$ with $n > 0$.

3.2   Linear iCol

In the following we analyze Linear iCol in a similar way as we did for Linear Col [8]. The only difference is that both players may use both colors (as long as they respect the Col-condition), which makes the analysis longer. On the other hand it suffices to only consider the options of one player, since the other player has exactly the same options by the impartial nature of the game; this shortens the analysis.

After coloring a node the Col-condition imposes that its neighbors never may receive the same color. To indicate this we may tint an empty neighbor of a black-colored node white, to show that such a node may only be colored White in the future. Similarly, an empty neighbor of a white node is tinted black. If an empty node receives both a black and a white tint, it means that this node may not be colored anymore at all.

We use the following notation for this: **B** for a black-colored square, **W** for a white-colored square, **b** for a black-tinted square, **w** for a white-tinted square, **x** for an empty square that can no longer be colored by either player, and **o** for an empty square that still can be colored by either player. For brevity we omit all edges. Our result is stated in Theorem 3.

Theorem 3. *Empty Linear iCol chains have CGT value 0 for even length and ∗ for odd length.*

Proof. Again when a node is colored (and the neighbors have been updated), it may be removed from the graph, which accordingly splits. It splits also when an empty node can not be colored, since this node may also be removed. Therefore the CGT value for a Linear iCol chain can be determined by the values of shorter subchains, in which only end nodes are possibly tinted. For the chains **o**···**o**, **b**···**b**, and **w**···**w** we consider only options in the left half of the chain, for other chains we consider all options. Of course symmetric chains like **b**···**o**, and **o**···**b** have the same values, just as swapping **b**s and **w**s in a chain (yielding the negation of the CGT value, which for nimbers has no effect). **0** denotes the Linear iCol chain of zero length, of course having value 0.

The analyses below always proceed in (at most) five steps: 1) determine the options of the chain; 2) simplify the options by removing colored and uncolorable nodes; 3) replace the options by their CGT values; 4) remove dominated options; and 5) determine the CGT value of the original chain.

Base Cases: $1 \times n$ chains with $n \leq 4$ have the following values:

$n = 1$: $\mathbf{o} = \{\mathbf{B}, \mathbf{W}\} = \{0, 0\} = \{0, 0\} = \{0\} = *$; $\mathbf{b} = \{\mathbf{B}\} = \{0\} = \{0\} = *$; similarly $\mathbf{w} = *$.

$n = 2$: $\mathbf{oo} = \{\mathbf{Bw}, \mathbf{Wb}\} = \{\mathbf{w}, \mathbf{b}\} = \{*, *\} = \{*\} = 0$; $\mathbf{bo} = \{\mathbf{Bw}, \mathbf{xB}, \mathbf{bW}\} = \{\mathbf{w}, 0, \mathbf{b}\} = \{*, 0, *\} = \{0, *\} = *2$; similarly, $\mathbf{ob} = \mathbf{wo} = \mathbf{ow} = *2$; $\mathbf{bb} = \{\mathbf{Bx}\} = \{0\} = \{0\} = *$; similarly $\mathbf{ww} = *$; $\mathbf{bw} = \{\mathbf{Bw}, \mathbf{bW}\} = \{\mathbf{w}, \mathbf{b}\} = \{*, *\} = \{*\} = 0$; similarly $\mathbf{wb} = 0$.

$n = 3$: $\mathbf{ooo} = \{\mathbf{Bwo}, \mathbf{wBw}, \mathbf{Wbo}, \mathbf{bWb}\} = \{\mathbf{wo}, \mathbf{w} + \mathbf{w}, \mathbf{bo}, \mathbf{b} + \mathbf{b}\} = \{*2, 0, *2, 0\} = \{0, *2\} = *$; $\mathbf{boo} = \{\mathbf{Bwo}, \mathbf{xBw}, \mathbf{bwB}, \mathbf{bWb}, \mathbf{bbW}\} = \{\mathbf{wo}, \mathbf{w}, \mathbf{bw}, \mathbf{b} + \mathbf{b}, \mathbf{bb}\} = \{*2, *, 0, 0, *\} = \{0, *, *2\} = *3$; similarly $\mathbf{oob} = \mathbf{woo} = \mathbf{oow} = *3$; $\mathbf{bob} = \{\mathbf{Bwb}, \mathbf{xBx}, \mathbf{bWb}\} = \{\mathbf{wb}, 0, \mathbf{b} + \mathbf{b}\} = \{0, 0, 0\} = \{0\} = *$; similarly $\mathbf{wow} = *$; $\mathbf{bow} = \{\mathbf{Bww}, \mathbf{xBw}, \mathbf{bWx}, \mathbf{bbW}\} = \{\mathbf{ww}, \mathbf{w}, \mathbf{b}, \mathbf{bb}\} = \{*, *, *, *\} = \{*\} = 0$; similarly $\mathbf{wob} = 0$.

$n = 4$: $\mathbf{oooo} = \{\mathbf{Bwoo}, \mathbf{wBwo}, \mathbf{Wboo}, \mathbf{bWbo}\} = \{\mathbf{woo}, \mathbf{w} + \mathbf{wo}, \mathbf{boo}, \mathbf{b} + \mathbf{bo}\} = \{*3, *3, *3, *3\} = \{*3\} = 0$; $\mathbf{booo} = \{\mathbf{Bwoo}, \mathbf{xBwo}, \mathbf{bwBw}, \mathbf{bowB}, \mathbf{bWbo}, \mathbf{bbWb}, \mathbf{bobW}\} = \{\mathbf{woo}, \mathbf{wo}, \mathbf{bw} + \mathbf{w}, \mathbf{bow}, \mathbf{b} + \mathbf{bo}, \mathbf{bb} + \mathbf{b}, \mathbf{bob}\} = \{*3, *2, *, 0, *3, 0, *\} = \{0, *, *2, *3\} = *4$; similarly $\mathbf{ooob} = \mathbf{wooo} = \mathbf{ooow} = *4$; $\mathbf{boob} = \{\mathbf{Bwob}, \mathbf{xBwb}, \mathbf{bWbb}\} = \{\mathbf{wob}, \mathbf{wb}, \mathbf{b} + \mathbf{bb}\} = \{0, 0, 0\} = \{0\} = *$;

similarly **woow** $=$ *; **boow** $=$ $\{$**Bwow, xBww, bwBw, bWbw, bbWx, bobW**$\}$ $=$ $\{$wow, ww, bw $+$ w, b $+$ bw, bb, bob$\}$ $=$ $\{*, *, *, *, *, *\} = \{*\} = 0$; similarly **woob** $= 0$.

So for $1 \leq n \leq 4$ we have

$$\mathbf{o} = \mathbf{b} = \mathbf{w} \qquad\qquad\qquad = *$$

$$\mathbf{o} \cdots \mathbf{o} \qquad\qquad = \begin{cases} 0 & \text{if } n \text{ is even} \\ * & \text{if } n \text{ is odd} \end{cases} \tag{1}$$

$$\mathbf{b} \cdots \mathbf{o} = \mathbf{w} \cdots \mathbf{o} = \mathbf{o} \cdots \mathbf{b} = \mathbf{o} \cdots \mathbf{w} = *n$$

$$\mathbf{b} \cdots \mathbf{b} = \mathbf{w} \cdots \mathbf{w} \qquad\qquad\quad = *$$

$$\mathbf{b} \cdots \mathbf{w} = \mathbf{w} \cdots \mathbf{b} \qquad\qquad\quad = 0$$

Induction hypothesis: suppose Eq. (1) holds for chains of length up to $k - 1$. **Induction steps:** consider a chain of length $k \geq 5$. We then have the following subcases, where a '$\cdots$' now indicates a sequence of nodes **o**, not of arbitrary length, but the length needed to have a complete chain of length k. For entries with chains '$\cdots$' at both sides of the colored square a range of possible entries is meant such that all combinations of left and right lengths are included with always a total length of k.

$\mathbf{o} \cdots \mathbf{o} = \{$**Bw**$\cdots$**o, wBw**$\cdots$**o, o**$\cdots$**wBw**$\cdots$**o**$\} = \{w\cdots$o, w $+$ w$\cdots$o, o$\cdots$ w $+$ w$\cdots$o$\} = \{*(k - 1), * + *(k - 2), \ldots, *(k - 2) + *, *(k - 1)\}$ (and similar for the first player using **W**, with the same values). For even k we see that every option is either an odd nimber or the Nim-sum of an odd plus even nimber, which is an odd nimber. Therefore the value of $\mathbf{o} \cdots \mathbf{o}$ for even k is 0 (being the $Mex()$ of all-odd nimbers). For odd k we see that every option is either an even nimber or the Nim-sum of two odd nimbers (which is an even nimber). This includes 0, namely when the middle node is colored (black or white), since both subchains are equal then. Therefore the value of $\mathbf{o} \cdots \mathbf{o}$ for odd k is * (being the $Mex()$ of all-even nimbers including 0).

$\mathbf{b} \cdots \mathbf{o} = \{$**Bw**$\cdots$**o, xBw**$\cdots$**o, b**$\cdots$**wBw**$\cdots$**o, b**$\cdots$**wBw, b**$\cdots$**wB, bWb**$\cdots$ **o, b**$\cdots$**bWb**$\cdots$**o, b**$\cdots$**bWb, b**$\cdots$**bW**$\} = \{$w$\cdots$o, w$\cdots$o, b$\cdots$w $+$ w$\cdots$o, b$\cdots$w $+$ w, b$\cdots$w, b $+$ b$\cdots$o, b$\cdots$b $+$ b$\cdots$o, b$\cdots$b $+$ b, b$\cdots$b$\} = \{*(k - 1), *(k - 2), 0 + *(k - 3), \ldots, 0 + *, 0 + 0, * + *(k - 2), * + *(k - 3), \ldots, * + *, *\} = \{0, \ldots, *(k - 1)\} = *k$. Similarly $\mathbf{o} \cdots \mathbf{b} = \mathbf{w} \cdots \mathbf{o} = \mathbf{o} \cdots \mathbf{w} = *k$.

$\mathbf{b} \cdots \mathbf{b} = \{$**Bw**$\cdots$**b, xBw**$\cdots$**b, b**$\cdots$**wBw**$\cdots$**b, bWb**$\cdots$**b, b**$\cdots$**bWb**$\cdots$**b**$\} = \{w\cdots$b, w$\cdots$b, b$\cdots$w $+$ w$\cdots$b, b $+$ b$\cdots$b, b$\cdots$b $+$ b$\cdots$b$\} = \{0, 0, \ldots, 0, * + *, \ldots, * + *\} = \{0\} = *$. Similarly $\mathbf{w} \cdots \mathbf{w} = *$.

$\mathbf{b} \cdots \mathbf{w} = \{$**Bw**$\cdots$**w, xBw**$\cdots$**w, b**$\cdots$**wBw**$\cdots$**w, b**$\cdots$**wBw, bWb**$\cdots$**w, b**$\cdots$ **bWb**$\cdots$**w, b**$\cdots$**bWx, b**$\cdots$**bW**$\} = \{$w$\cdots$w, w$\cdots$w, b$\cdots$w $+$ w$\cdots$ w, b$\cdots$w $+$ w, b $+$ b$\cdots$w, b$\cdots$b $+$ b$\cdots$w, b$\cdots$b, b$\cdots$b$\} = \{*, *, 0 + *, 0 + *, * + 0, * + 0, *, *\} = \{*\}$. So $\mathbf{b} \cdots \mathbf{w} = 0$. Similarly $\mathbf{w} \cdots \mathbf{b} = 0$.

This means that based on the assumption that Eq. (1) holds for chain length up to $k - 1$ it follows that it holds for chain length k. Combined with the base cases, Eq. (1) consequently holds for arbitrary chain lengths. □

Concludingly, in Linear iCol even-length empty chains have CGT value 0 (second-player wins) and odd-length empty chains have CGT value $*$ (first-player wins). In the latter case the first-player must color the middle square (either black or white).

3.3 iSnort on Rectangular Boards

For $m \times n$ iSnort boards with m and/or n even the second player always can win. This is proven in Theorem 4.

Theorem 4. *All empty $m \times n$ iSnort boards with m and/or n even are second-player wins and thus have CGT value 0.*

Proof. The second player can again use a copy-strategy as in iCol, but always using the same color as the previous move (the center-same strategy). Therefore, after every second-player's move the board is center-symmetric with same colors. Consequently, the second player makes the last move and wins. □

When the first player just sticks to using one color, the game ends after the full board has been filled with one color. Example iSnort games on the 4×4 and 4×5 boards where Right uses this winning strategy are shown in Fig. 3.

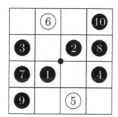 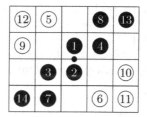

Fig. 3. Example iSnort games on the 4×4 and 4×5 boards won by Right.

The left diagram shows an iSnort game on an even $\times$ even board, the right diagram on an even $\times$ odd board. Right has chosen to always use the center-same strategy, guaranteeing the win. Note that for iSnort on an even $\times$ even board an alternative winning strategy for the second player would be to use the center-opp strategy. For odd $\times$ even and even $\times$ odd boards this strategy is not possible, since it might violate the Snort-condition.

Analogously with iCol, for $m \times n$ iSnort boards with m and n odd the first player always can win. This is proven in Theorem 5.

Theorem 5. *All empty $m \times n$ iSnort boards with m and n odd are first-player wins.*

Proof. Like in iCol, the first player can easily win by first coloring the center square arbitrarily, followed by using the center-same strategy. Since the number of empty squares after the first move is even, this guarantees the first player to make the last move and win. □

Again, we only know that such boards have fuzzy values, but since all values in iSnort must be nimbers, we know that the values of odd × odd boards have nimber values $*n$ with $n > 0$.

An example game where the first player uses this strategy to win the game is given in Fig. 4.

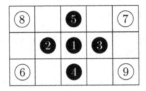

Fig. 4. Example iSnort game on the 3 × 5 board won by Left.

3.4   Linear iSnort

For Linear iSnort the situation is quite similar as for iCol, namely second-player wins (CGT value 0) for even-length chains, and first-player wins (CGT values $*n$ with $n > 0$) for odd-length chains. For iCol we found that all odd-length chains have value $*$ (Theorem 3). To prove that this is also the case for iSnort we analyse Linear iSnort in a similar way as we did for Linear iCol, using the same notations. Again both players may use both colors, but now the moves must respect the Snort-condition. Our result is stated in Theorem 6.

Theorem 6. *Empty Linear iSnort chains have CGT value 0 for even length and $*$ for odd length.*

Proof. This proof is analogous as the proof of Theorem 3, of course except the Snort-condition on neighboring squares instead of the Col-condition. We just give the main parts of the proof and leave the complete analysis as an exercise.

Base Cases: $1 \times n$ chains with $n \leq 4$ have the following values:

$n = 1$: $\mathbf{o} = \mathbf{b} = \mathbf{w} = \{0\} = *$.
$n = 2$: $\mathbf{oo} = \{*\} = 0$; $\mathbf{bo} = \mathbf{ob} = \mathbf{wo} = \mathbf{ow} = \{0, *\} = *2$; $\mathbf{bb} = \mathbf{ww} = \{*\} = 0$;
    $\mathbf{bw} = \mathbf{wb} = \{0\} = *$.
$n = 3$: $\mathbf{ooo} = \{0, *2\} = *$; $\mathbf{boo} = \mathbf{oob} = \mathbf{woo} = \mathbf{oow} = \{0, *, *2\} = *3$;
    $\mathbf{bob} = \mathbf{wow} = \{0\} = *$; $\mathbf{bow} = \mathbf{wob} = \{*\} = 0$.
$n = 4$: $\mathbf{oooo} = \{*3\} = 0$; $\mathbf{booo} = \mathbf{ooob} = \mathbf{wooo} = \mathbf{ooow} = \{0, *, *2, *3\} = *4$;
    $\mathbf{boob} = \mathbf{woow} = \{*\} = 0$; $\mathbf{boow} = \mathbf{woob} = \{0\} = *$.

So for $1 \leq n \leq 4$ we have

$$\mathbf{o} = \mathbf{b} = \mathbf{w} \qquad\qquad\qquad = *$$

$$\mathbf{o} \cdots \mathbf{o} \qquad\qquad\qquad = \begin{cases} 0 & \text{if } n \text{ is even} \\ * & \text{if } n \text{ is odd} \end{cases}$$

$$\mathbf{b} \cdots \mathbf{o} = \mathbf{w} \cdots \mathbf{o} = \mathbf{o} \cdots \mathbf{b} = \mathbf{o} \cdots \mathbf{w} = *n$$

$$\mathbf{b} \cdots \mathbf{b} = \mathbf{w} \cdots \mathbf{w} \qquad\qquad = \begin{cases} 0 & \text{if } n \text{ is even} \\ * & \text{if } n \text{ is odd} \end{cases} \qquad (2)$$

$$\mathbf{b} \cdots \mathbf{w} = \mathbf{w} \cdots \mathbf{b} \qquad\qquad = \begin{cases} * & \text{if } n \text{ is even} \\ 0 & \text{if } n \text{ is odd} \end{cases}$$

Induction hypothesis: suppose Eq. (2) holds for chains of length up to $k - 1$.
Induction steps: consider a chain of length $k \geq 5$. We then have the following subcases.

$\mathbf{o} \cdots \mathbf{o} = \{*(k - 1), * + *(k - 2), \ldots, *(k - 2) + *, *(k - 1)\}$ for using Black (and similar for the White options, with the same values). The value of $\mathbf{o} \cdots \mathbf{o}$ for even k is 0, for odd k it is $*$.

$\mathbf{b} \cdots \mathbf{o}$: We have to distinguish cases where the chain has odd or even length, and whether the player uses Black or White as color.
We first consider even k. Using Black the sums of the left and right subchains are $*(k - 1)$, $*(k - 1)$, $*(k - 3)$, $*(k - 3)$, etc., i.e. all odd nimbers from $*$ to $*(k - 1)$. Using White the sums of the left and right subchains are $*(k - 2)$, $*(k - 4)$, $*(k - 4)$, etc., i.e. all even nimbers from 0 to $*(k - 2)$. Taking all options together we see that all nimbers from 0 to $*(k - 1)$ are included, which means that the value of the total chain for even k is $*k$.
For odd k the analysis is quite similar, this time leading to the series $*(k - 1)$, $*(k - 3)$, $*(k - 3)$, $*(k - 5)$, $*(k - 5)$, etc. for the options using Black, including all even nimbers from 0 to $*(k - 1)$; for the options using White we obtain the series $*(k - 2)$, $*(k - 2)$, $*(k - 4)$, $*(k - 4)$, etc., again including all odd nimbers from $*$ to $*(k - 2)$. Taking all options together we see that again all nimbers from 0 to $*(k - 1)$ are included, which means that the value of the total chain for odd k is also $*k$.

$\mathbf{b} \cdots \mathbf{b}$: We differentiate between the Black and White options.
For options using Black the chain is split in two parts, with either value 0 (even-length $\mathbf{b} \cdots \mathbf{b}$ subchains) or $*$ (odd-length $\mathbf{b} \cdots \mathbf{b}$ subchains). For options using White the chain is split in two parts, with either value $*$ (even-length $\mathbf{b} \cdots \mathbf{w}$ subchains) or 0 (odd-length $\mathbf{b} \cdots \mathbf{w}$ subchains). In either case, for even k the sum of the two splits has odd length, with sum value $*$, which means the original chain has value 0; for odd k the sum of the two splits has even length, with sum value 0, which means the original chain has value $*$. Similarly, a chain $\mathbf{w} \cdots \mathbf{w}$ with length k also has value 0 if k is even, and value $*$ if k is odd.

b⋯**w**: We again differentiate between the Black and White options. For options using Black the chain is split into a **b**⋯**b** subchain (odd-length *, even-length 0) and a **b**⋯**w** subchain (odd-length 0, even-length *). For options using White the chain is split into a **b**⋯**w** subchain (odd-length 0, even-length *) and a **w**⋯**w** subchain (odd-length *, even-length 0). In either case, for even k the sum of the two splits has sum value 0, which means the original chain has value *; for odd k the sum of the two splits has sum value *, which means the original chain has value 0. Similarly, a chain **w**⋯**b** with length k also has value * if k is even, and value 0 if k is odd.

This means that based on the assumption that Eq. (2) holds for chain length up to $k - 1$ it follows that it holds for chain length k. Combined with the base cases, Eq. (2) consequently holds for arbitrary length chains.   □

Concludingly, like in Linear iCol, in Linear iSnort even-length empty chains have CGT value 0 (second-player wins) and odd-length empty chains have CGT value * (first-player wins). In the latter case the first-player must color the middle square (either black or white).

4   Conclusions and Future Research

We summarize our main results in Table 1, where we give for all board types the corresponding outcome class. Note that the results for even × odd boards are equal to their equivalent odd × even boards obtained by a 90° rotation.

Table 1. Outcome classes for iCol and iSnort on boards of different types.

Game	even × even	odd × even	odd × odd
iCol	$\mathcal{P}$	$\mathcal{P}$	$\mathcal{N}$
iSnort	$\mathcal{P}$	$\mathcal{P}$	$\mathcal{N}$

This table shows that iCol and iSnort on rectangular boards are solved games, though their values on odd × odd boards can vary (nimbers *n with $n > 0$). For Linear iCol and iSnort we do have precise CGT values, given in Table 2.

Table 2. CGT values for Linear iCol and iSnort on chains of length n.

Game	even n	odd n
Linear iCol	0	*
Linear iSnort	0	*

All values in this paper were checked with the CGSUITE [6] system and fully agree with our findings.[1] For future research we are also interested in results for iCol and iSnort played on other graphs than rectangular boards.

References

1. Albert, M.H., Nowakowski, R.J., Wolfe, D.: Lessons in Play: An Introduction to Combinatorial Game Theory. A K Peters, Wellesley (2007)
2. Berlekamp, E.R., Conway, J.H., Guy, R.K.: Winning Ways for your Mathematical Plays. Academic Press, London (1982). 2nd edition, in four volumes: vol. 1 (2001), vols. 2, 3 (2003), vol. 4 (2004). A K Peters, Wellesley
3. Bouton, C.I.: Nim, a game with a complete mathematical theory. Ann. Math. **3**, 35–39 (1902)
4. Conway, J.H.: On Numbers and Games. Academic Press, London (1976)
5. Siegel, A.N.: Combinatorial Game Theory, Vol. 146 in Graduate Studies in Mathematics. American Mathematical Society (2013)
6. Siegel, A.N.: Combinatorial game suite: a computer algebra system for research in combinatorial game theory (2020). http://cgsuite.sourceforge.net/
7. Silverman, D.L.: Your Move. McGraw-Hill, New York (1971). Revised and reprinted as Your Move: Logic, Math and Word Puzzles for Enthusiasts. Dover Publ. Inc, New York (1991)
8. Uiterwijk, J.W.H.M.: Solving bicoloring-graph games on rectangular boards - part 1: partisan Col and Snort. In: Browne, C., et al. (eds.) ACG 2021. LNCS, vol. 13262, pp. 96–106. Springer, Cham (2022)

---

[1] The CGSUITE code for the board implementations of iCol and iSnort were simple adaptations of the ones for Col and Snort as used in [8].

BoxOff is NP-Complete

Ryan Hayward$^{(\boxtimes)}$, Robert Hearn, and Mahya Jamshidian

University of Alberta, Edmonton, AB T6G 2R3, Canada
{hayward,mjamshidian}@ualberta.ca

Abstract. BoxOff is a one-player game invented by Steven Meyers: a rectangular board is covered with colored stones; a legal move is to remove two same-colored stones from opposite corners of an otherwise empty rectangle. Removing all stones wins the game. We show that it can be hard to determine whether a BoxOff puzzle is winnable: by reducing from Boolean Satisfiability, we show that BoxOff is NP-complete, even when only four colors are used.

Keywords: BoxOff puzzle · NP-Complete · Satisfiability

1 Introduction

In 2013 in *Games Magazine*, Steven Meyers introduced his new solitaire game BoxOff: see the article by Kerry Handscomb in *Abstract Games Magazine* for a colorful introduction [3,6]. The board has a rectangular grid; each board cell is empty or has a colored stone. On a move, the player removes two stones of the same color that lie on opposite corners of an otherwise empty rectangle. The player wins by clearing the board. We are interested in this decision question: given a BoxOff puzzle, is it solvable, i.e. can the player win? Consider Fig. 1. The left puzzle is solvable, e.g. remove {a1,b1}, then {b2,c2}, then {a2,c1}. The right puzzle is not solvable: each of {b2,c1}, {b1,c2} must be removed before the other, which is impossible. To learn the basics of BoxOff strategy, see Handscomb's article.

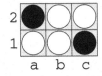
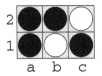

Fig. 1. Two 2-color BoxOff puzzles (left is solvable, right is not).

Browne and Maire investigated the complexity of BoxOff [1]. They gave a Monte Carlo analysis of random play, described a polytime algorithm to solve one-column k-color BoxOff, and noted that otherwise the puzzle's complexity was unknown. We will show that 4-color BoxOff is NP-complete, resolving this open question.

© Springer Nature Switzerland AG 2022
C. Browne et al. (Eds.): ACG 2021, LNCS 13262, pp. 118–127, 2022.
https://doi.org/10.1007/978-3-031-11488-5_11

2   Reduction Overview

In the usual way [2,5], we will show that 4-color BoxOff is NP-hard by reducing from 3-CNF-Satisfiability (3-SAT). Given a 3-SAT formula, we construct a Box-Off puzzle that is solvable if and only if the formula is satisfiable. The BoxOff puzzle models a Boolean circuit, formed from gadgets (independent sub-puzzles) that simulate Boolean variables, AND and OR gadgets, fanout gadgets, color-switch gadgets, and turn gadgets. In order to ensure that gadgets influence each other only with respect to the flow of the circuit, we insulate the gadgets by placing each inside one cell of a larger overlay grid.

We will show that the input formula is satisfiable if and only if the Box-Off puzzle can be cleared by a two-phase process, starting with a multiple-source/single-sink flow from each variable to a single cell indicating that the complete formula is satisfied, followed by a cleanup phase that erases all overlay stones and any remaining gadget stones.

3   4-Color BoxOff is NP-Complete

Here we give our reduction, and also show an example: the BoxOff puzzle corresponding to $(x \vee y) \wedge (x \vee \bar{y}) \wedge (\bar{x} \vee y)$.

We transform the input formula into a puzzle by connecting the literal gadgets to appropriate OR gadgets and thence AND gadgets, using the wiring (turn, fanout, and color-change) gadgets. We do not need a crossover gadget: there is only empty space between the stones that are paired, and the overlay grid keeps any gadgets from interacting that are not directly paired in a row or column.[1]

3.1   Overlay Stones and Gadget Stones

We use two stone colors (black and white) for the overlay grid. Within each row or column that includes an overlay stone, the stones alternate colors. We use two other colors (red and blue) for our gadgets. Each gadget fits in a bounded grid called a *container*, defined as the empty rectangular regions within the overlay grid. We align gadgets so that one gadget's output is on the same line (horizontal or vertical) as the next gadget's input. At least one overlay stone separates any pair of stones not in gadgets in the same row or column of the overlay grid, so they can never interact. We only put two gadgets in the same row or column when they are connected. Figure 2 is an example of the overlay grid.

---

[1] Reducing instead from Planar 3-SAT would not help us avoid the need for a crossover, if crossing signals were not trivial in this setting. The reason is that we need to connect the clauses together to produce a single output signal, a situation that is common in SAT reductions. But Planar 3-SAT only applies when the graph connecting the variables to their clauses is planar; it does not allow us to further connect the clauses. Instead, we would then have reduced from Bounded One-Player Constraint Logic [4], which solves this problem.

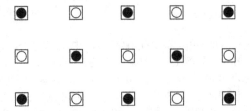

Fig. 2. The overlay black and white stones, demarking the containers within. The gadget stones are located in these containers, to prevent unwanted interactions.

3.2 Signals

Signals are propagated whenever an *output* stone from one gadget is paired with an *input* stone in another gadget, removing both. We then say that the output and corresponding input have been *activated*. The activation of its inputs (if any) is what allows a gadget to activate its output(s).

3.3 Variable Gadget

A *variable* gadget consists of a single output stone, connecting to inputs in two other gadgets. See Fig. 3(a). Each variable corresponds to a switch, where the player can set the output signal to be true or false. So each gadget has two possible output directions, indicated in the figure by arrows. A satisfying assignment corresponds to setting the appropriate output signal (true or false) for each variable gadget.

3.4 Wiring Gadgets

Lemma 1. *The* turn *gadget rotates its input signal by 90 °C.*

Proof. When the input signal is available, the red stone can be matched and removed, allowing the blue stone to propagate the output signal.

Lemma 2. *The* fanout *gadget splits one input signal into two.*

Proof. An active input signal matches the middle blue stone, allowing the lower and the upper red stone to be removed. The remaining blue stones can leave the gadget as active signals.

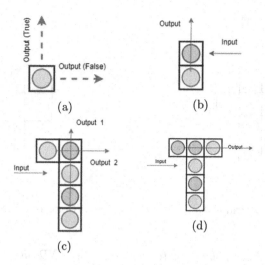

Fig. 3. Gadgets: (a) Variable (b) Turn (c) Fanout (d) Color-change. (Color figure online)

Lemma 3. *The* color-change *gadget turns a blue input signal into a red output signal.*

Proof. An active input signal matches the middle blue stone, allowing the upper-middle and lower red stone to match and be removed. The remaining blue stones match and the upper-left red stone serves as an output signal.

There are two ways to change colors: to maintain the direction of the input signal, use the color-change gadget; to change the direction of the signal, use the turn gadget. (The color-change gadget is actually not necessary: we could just combine a fanout gadget and turn gadget instead. However, the color-change gadget makes our transformations slightly more compact.) (Fig. 4).

3.5   Logical Gadgets

Lemma 4. *The* OR *gadget functions as a logical OR operator.*

Proof. Case 1) Both inputs are active: Both upper-level blue stones are matched with input signals. Then the upper red stone is matched with either of the lower red stones. Then the lower blue stone can leave the gadget as a active signal. The two remaining red stones will match once the output signal is activated, i.e. once the lower blue stone disappears.

Case 2) Only input 1 is active: The upper-right blue stone matches input signal 1. Then the upper red stone matches the right red stone, allowing the output stone to leave the gadget as an active signal.

Case 3) Only input 2 is active: Similar to the previous case.

Case 4) Neither input is active: No stones are matched: the output signal cannot leave the gadget, so is inactive.

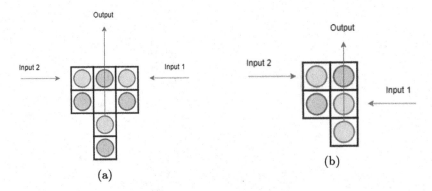

Fig. 4. Logical gadgets: (a) OR, (b) AND. (Color figure online)

Lemma 5. *The* AND *gadget functions as a logical AND operator.*

Proof. Case 1) Both inputs are active: The upper and middle blue stones match the inputs. Then the red stones match, so the output blue stone is an active signal.

Case 2) At most one input is active: At least one blue stone remains, preventing the red stones from matching, so the output signal is inactive.

3.6   Gadget Interactions

We have already shown that gadgets not in the same row or column cannot interact. But we also need to show that gadgets that are in the same row or column can only interact as intended. Ideally this would mean that no pairs of stones can be removed other than in the valid propagation of signals. In fact, a slightly weaker condition suffices. First, we only connect non-turn gadgets together via intervening turns, never directly. Then we only need to analyze connections involving turns. But a turn has only two stones: the input stone is the one intended to be matched, and can only be paired with its intended partner; the output stone should instead be used to propagate the signal onward. If instead the output stone is matched with a stone in another gadget, then the signal simply doesn't propagate, and no incorrect solution of the puzzle is thus enabled. But what about the loss of the stone that inappropriately paired with the turn's output stone? This can only be a stone in the turn's input gadget. If that gadget has a single output, then any change in its properties is irrelevant, because the output can't propagate further. The only other case is that the input gadget is a fanout. We need to verify that if a fanout stone inappropriately pairs with an output turn gadget's output stone, this does not then enable the other output to activate. By inspection, this is the case.

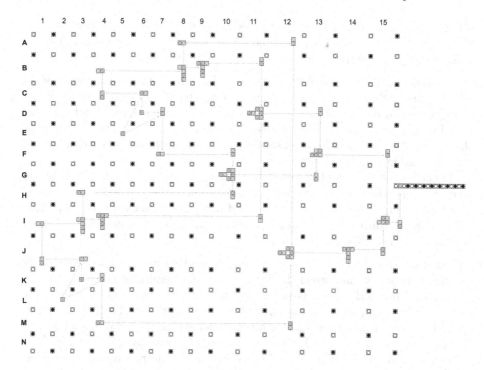

Fig. 5. Transformation of $(x \vee y) \wedge (x \vee \bar{y}) \wedge (\bar{x} \vee y)$. Gadgets for x,y in cells K3,D6, auxiliary stones at L2,E5. OR at G10 is $(x \vee \bar{y})$, OR at D11 is $(x \vee y)$, OR at J12 is $(\bar{x} \vee y)$. AND at F13 is $((x \vee y) \wedge (x \vee \bar{y}))$, AND at I15 is $(x \vee y) \wedge (x \vee \bar{y}) \wedge (\bar{x} \vee y)$.

3.7 The Reduction

The main result of our paper is the following theorem. The example transformation shown in Fig. 5 may serve as a reference here; it is analyzed explicitly in Sect. 3.8. (Note: for compactness, in the example we relax the rule about only connecting non-turn gadgets via turns; in this instance no unwanted interactions arise.)

Theorem 1. *Deciding whether a BoxOff game with 4 colors is solvable is NP-Complete.*

Proof. Let ϕ be a 3-CNF formula with variables $x_1, \cdots, x_n$. Then we construct a corresponding BoxOff configuration as follows.

Variable Gadgets. There is one gadget for each variable x_j.

OR *gadgets.* Our OR gadgets take two inputs. Each clause has three literals, so we use two chained OR gadgets for each clause.

Fanout Gadgets. A variable can appear in more than one clause, so we might need to duplicate a variable's output signal to all instances of the corresponding literals. For this, we use fanout gadgets, possibly sending signals far around the grid, so we also need turn (and if necessary color-change) gadgets. A variable gadget's True (resp. False) output signal flows to OR gadgets in which the variable is a positive (resp. negated) literal.

AND *Gadgets:* We use the AND gadgets to logically combine the value of all clauses; again, this operator takes only two inputs, so we need to chain $k - 1$ AND gadgets to resolve a formula with k clauses. We can activate the final AND if and only if the formula is satisfiable. Notice that when at least one input to the AND gadgets is inactive, at least one AND gadget will have uncleared stones, and the BoxOff puzzle will not be solvable.

In the rest of this proof, we want to show that the BoxOff puzzle is solvable when the input formula is satisfiable. Hence, assume the selected assignment is satisfying; then, the construction described allows us to activate the final AND gadget, by activating the variable gadgets appropriately and propagating all signals where possible.

Cleaning Up Overlay Grid: In order to be able to clean up the board, we modify our construction so that when the final AND gadget emits an active signal, we can remove the overlay grid. We route this active signal to a special middle row of our construction, where to the right of the final overlay-grid stone, say black, we add a red stone and then another black stone, and then alternating white and black stones, enough to match the middle row of the overlay grid. See Fig. 6(a).

In Fig. 6(b), notice that the red stone in the middle row matches the final AND active signal, so the middle-row black stones adjacent to the red stone will match, allowing the middle row to disappear completely, which will then allow each column to disappear, as we construct the overlay grid so that within each row and column, the black and white stones alternate, and the total number of overlay stones in each column is odd.

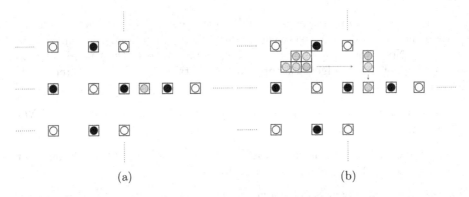

(a)   (b)

Fig. 6. (a) Rightmost edge of the grid, showing the middle row. (b) Rightmost edge of the grid, with middle row and final AND gadget. (Color figure online)

Cleaning Up Remaining Stones: Depending on the particular Boolean assignment selected for our formula, each gadget (except for the AND gadgets) might have remaining stones, either because the signal never reached the gadget, or the signal reached the gadget but not all stones were removed when the signal flowed through. We now show how to add extra stones to variable gadgets so that a final cleanup is possible whenever the formula is satisfied.

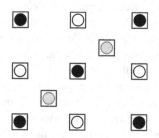

Fig. 7. The modified variable gadget has an extra stone located in a diagonally adjacent cell.

The idea is that the extra stone can activate the True/False signal not assigned to the variable. This auxiliary stone must not interact with any gadget until the overlay grid has vanished; we place it in a diagonally adjacent cell, as shown in Fig. 7.

We constructed our gadgets so that they normally receive inputs in a straight line, but the auxiliary stone will not be in a line with the variable's output turn gadgets. To address this, we orient those turn gadgets such that the auxiliary stone can still pair with them, as shown in Fig. 8.

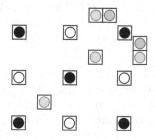

Fig. 8. The modified variable gadget, showing the connecting turn gadgets and the auxiliary stone to be used during the clean-up phase.

Once all overlay stones and the satisfied parts of the gadgets are gone, the auxiliary stone can be matched with the turn at the output of the unchosen path and clear the following stones.

Once the auxiliary stones for each unactivated literal path have cleared, each instance of each gadget has disappeared: variable and auxiliary gadgets have cleared; ANDs have cleared; ORs all have both inputs activated (because each traces back to a literal, or to another OR that has activated); fanouts split variable outputs, and since all variable outputs are active, fanouts have cleared; turns occur only on pathways connecting the above and so have cleared. So, with a satisfying assignment, we can clear the entire board. If there is no satisfying assignment, then the final AND gadget is not cleared by signal propagation, and overlay grid ensures that this gadget cannot be cleared in any other way.

Our reduction is polynomial: we have one AND or OR gadget per logical operator in the formula, and one variable or fanout per literal. We can place each variable, auxiliary variable, AND, OR, and fanout in its own column and row. If we allow an equal number of rows and columns for routing, we can connect any gadget output to any other gadget input with at most four turns and one color-change gadget. Thus the number of rows and columns needed is polynomial in the input size. The size of each container (i.e., the spacing of the overlay stones) also need only be polynomial in the input size: the gadgets might need to be positioned at various offsets within the containers to align inputs and outputs, but at worst the required space would be polynomial in the total number of gadgets, and all we actually need is that the logarithm of this spacing is polynomial.

Finally, BoxOff is clearly in NP: a solution is a list of pairings, which is of polynomial length, and can be easily verified. This completes the proof.

3.8 Example Transformation

Here we describe our example transformation of $(x \vee y) \wedge (x \vee \bar{y}) \wedge (\bar{x} \vee y)$ shown in Fig. 5. For the sake of simplicity, this example is in 2-CNF.

To begin, consider a trial assignment, say x and y both false. Since x is false, we follow the horizontal-right output signal from variable gadget x (at K3), which then matches the red stone at $K4$. The signal then proceeds to the lower input of the OR gadget at $J12$, then to the color-change gadget at $J14$, through a turn gadget at $J15$, and reaches the lower-level input of the AND gadget at $I15$.

Meanwhile, starting from variable gadget y at $D6$, output proceeds to the turn at $D7$, follows arrows to the upper input of the OR gadget at $G10$, and can leave the gadget to reach the lower input of the AND gadget at $F13$.

Notice that these two signals cannot move further, since the D11 OR gadget has no active input. Thus neither AND gadget at F13,I15 has an active upper input, so each is inactive. So our current assignment fails.

Next consider the satisfying assignment with both x, y true. Now the x variable stone at K3 activates the turn at J3 and the signal continues to the I3 fanout gadget, whose vertical output continues to the lower input of the G10 OR. The horizontal fanout output activates the I4 color-change gadget at $I4$ and then the D11 OR lower input.

Also, the y variable at $D6$ matches the red turn stone at $C6$. The output of the turn gadget matches the input of the B8 fanout, whose vertical output activates the upper input of the J12 OR. The vertical fanout output matches the upper input of the D11 OR.

Now all OR gadgets have active outputs, so the two AND gadgets have both inputs active, so both have active outputs. The output of the I15 AND matches the red stone outside of the overlay grid, so the middle black-white row clears, and each black-white column then clears.

After all black and white stones are gone, we are left with six turn gadgets, two blue stones from the OR gadgets J12 and G10, and the auxiliary stone for both x and y. The auxiliary x-stone at $L2$ matches the red stone at $K4$ and eventually clears out the remaining lower blue stone for the OR gadget at $J11$, and all turn gadgets in between. The auxiliary y-stone at $E5$ matches the D7 turn gadget and—the last step—the upper blue stone for the G10 OR, and all turn gadgets in between. Finally, the board is clear.

4   Conclusion

We showed that 4-color BoxOff is NP-complete, resolving an open question of Browne and Maire [1]. Our 3SAT reduction is straightforward, except for the extra machinery needed to clean up the board once the final AND activates. Our reduction requires at least four colors: two for the overlay grid and two for the gadgets. Might a different approach show hardness for three, or even two colors? (One color is trivial.) We conjecture that with three colors BoxOff is still hard, but with two it is polynomial. We encourage further work to clarify this fascinating boundary.

Acknowledgment. We would like to show our gratitude to Steven Meyers, the designer of the BoxOff game, for his careful review of the paper.

References

1. Browne, C., Maire, F.: Monte Carlo analysis of a puzzle game. In: Pfahringer, B., Renz, J. (eds.) AI 2015. LNCS (LNAI), vol. 9457, pp. 83–95. Springer, Cham (2015). https://doi.org/10.1007/978-3-319-26350-2_8
2. Cook, S.A.: The complexity of theorem-proving procedures. In: Harrison, M.A., Banerji, R.B., Ullman, J.D. (eds.) Proceedings of the 3rd Annual ACM Symposium on Theory of Computing, 3–5 May 1971, Shaker Heights, Ohio, USA, pp. 151–158. ACM (1971)
3. Handscomb, K.: Steven Meyers' BoxOff. Abstract Games Magazine, vol. 19, pp. 35–36 (2020)
4. Hearn, R.A., Demaine, E.D.: Games, Puzzles and Computation. A K Peters (2009)
5. Karp, R.M.: Reducibility among combinatorial problems. In: Miller, R.E., Thatcher, J.W. (eds.) Proceedings of a Symposium on the Complexity of Computer Computations, held 20–22 March 1972, at the IBM Thomas J. Watson Research Center, Yorktown Heights, New York, USA, pp. 85–103. The IBM Research Symposia Series. Plenum Press, New York (1972)
6. Meyers, S.: Boxoff game. Games Magazine, August 2013

Chess Patterns

Automatic Recognition of Similar Chess Motifs

Miha Bizjak and Matej Guid[✉]

Faculty of Computer and Information Science, University of Ljubljana, Ljubljana,
Slovenia
matej.guid@fri.uni-lj.si

Abstract. We present a novel method to find chess positions similar to
a given query position from a collection of chess games. We consider not
only the static similarity resulting from the arrangement of chess pieces,
but also the dynamic similarity involving the recognition of chess motifs
and the tactical, dynamic aspects of position similarity. By encoding
chess tactical problems as text documents, we use information retrieval
techniques to enable efficient approximate searches. We have also developed a method for automatically generating tactical puzzles from a collection of chess games. We have experimentally shown the importance of
including both static and dynamic features for successful recognition of
similar chess motifs. The experiments have clearly shown that dynamic
similarity plays a very important role in the evaluation of the similarity
of chess motifs by both the program and chess experts.

Keywords: Problem solving · Chess motifs · Automatic similarity
recognition

1 Introduction

The focus of this paper is on automatic retrieval of similar tactical problems from
a large collection of chess games. The term *tactic* is used in chess to describe
a series of moves that exploit a particular position on the board and allow the
player to gain material, gain a positional advantage, or even force a checkmate.

For chess players to progress, tactical problems are incredibly important.
Knowing tactical motifs helps them to recognise when a winning or drawing
combination might exist in a position. By solving tactical problems, chess players improve their tactical skills. It is not uncommon for games to be decided by
tactics, because even a single mistake gives the opportunity to use a tactic that
changes the outcome of the game. A large number of patterns or *tactical motifs*
have been defined in chess literature to help players discover tactical opportunities during a game [2]. The ability to recognise chess motifs during a game is
one of the key components of becoming a competent chess player [4].

In order to provide useful teaching material to their students, chess instruc-
tors often look for examples from actual games that exhibit relevant chess motifs.

© Springer Nature Switzerland AG 2022
C. Browne et al. (Eds.): ACG 2021, LNCS 13262, pp. 131–141, 2022.
https://doi.org/10.1007/978-3-031-11488-5_12

However, the human mind is not capable of sifting through thousands or even millions of games to find problems with similar chess motifs and similar solutions to those overlooked by students. Contextually similar chess positions could also be used to annotate chess games [6] and in intelligent tutoring systems [8].

Our research aims for the automatic retrieval of similar chess motifs for a given query position in a large collection of archived chess games. It needs to be both efficient in terms of speed and effective in terms of the quality/similarity of the retrieved results.

1.1 Related Work

There are a large number of possible positions in a game of chess, so a query-by-example [9] system that only looks for exact matches would not be effective. To mitigate the problem, the Chess Query Language (CQL) [3] allows searching for approximate matches of positions, but requires the user to define complex queries in the system-specific language. In addition, CQL works directly with PGN game archive files and checks each game sequentially, making it inefficient for querying larger databases.

To overcome these problems, an information retrieval-based approach was proposed, in which a textual representation is constructed for each board position and information retrieval (IR) methods are used to compute the similarity between chess positions [5]. Instead of creating a query manually, the user simply enters a FEN and the query encoding the features of the position is automatically created internally. Initially, a naïve coding was used, which only contained the positions of the individual pieces. Additional information about the mobility of the individual pieces and the structural relationships between the pieces improved the results.

Besides improving the above approach by introducing advanced static patterns such as pawn structures, we aim to develop a state-of-the-art chess engine for finding similar chess tactics that also takes into account the dynamic aspects of chess tactics. In terms of the progress of chess players, this dynamic aspect is much more relevant when considering a contextually similar tactical problem.

2 Domain Description

Figure 1 illustrates three chess tactics with the same chess motif: the white rook is sacrificed in the corner of the board and the black king must capture it, allowing the white queen to appear with check (it cannot be captured due to the activity of the white bishop along the long diagonal) and deliver checkmate on the next move. Note that the similarity is due to a dynamic aspect of the combination, which is based on the underlying chess motif. We use standard chess annotation to describe chess moves.

To illustrate the difference between *static* and *dynamic similarity*, consider the removal of the white rook on the h-file in each of the three positions in Fig. 1.

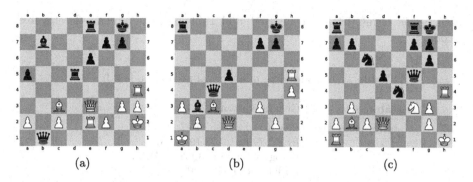

Fig. 1. Chess tactics with the same chess motif. In all three positions, White wins with 1.♖ h8+ ♚ xh8 2.♛ h6+ ♚ g8 3.♛ xg7 checkmate.

The new positions would be statically very similar, but the dynamic similarity would disappear and the tactic mentioned above would no longer apply.

We are particularly interested in detecting dynamic similarity, i.e. finding positions with similar motif(s) in the solution of the tactic. However, we also want to consider static similarity, i.e. finding problems with similar initial positions. Figure 1 also illustrates the behaviour of our program: when querying a chess tactic in (a), the program found two similar chess tactics, shown in (b) and (c).

3 Similarity Computation

To determine similarity between tactical problems we use an approach based on information retrieval. A set of features is computed from each problem's starting position and its solution move sequence. The features are then converted into textual terms, forming a document that represents the problem. A collection of documents is used to build an index, which can then be queried using the textual encoding of a new position to retrieve the most similar positions in the index.

The conversion from the two-dimensional structure given by the chess board to the set of independent textual terms allows us to make use of highly optimized and scalable text information retrieval systems, while still preserving important features of each position by encoding specific piece combinations and interactions between pieces on the board.

For the implementation of our system for indexing and retrieval of similar tactics we use the *Apache Lucene Core* library. Search results are ranked using the BM25 ranking function [7]. For our use case, the most important character-istics of BM25 are increased relative score contributions of less frequent terms using each term's inverse document frequency (IDF) and normalization of scores based on document length, favoring shorter documents over longer ones with a large number of irrelevant terms.

For each tactic, the input consists of a starting position in FEN format and a solution move sequence in algebraic notation. The solution can be provided with the position or calculated using a chess engine. Sections 3.1 and 3.2 describe the features and terms that are generated, and Table 1 shows an example of a text encoding.

Table 1. Text encoding of a tactical position in Fig. 1c.

Feature set	Generated terms
static_positions	Ra1 Kh1 Pa2 Bb2 Pc2 Qd2 Pg2 ...
	Rb1|0.89 Rc1|0.78 Rd1|0.67 ...
	B>pg7 Q>pd5 n>Qd2 n>Pg3 ...
	R<Kh1 R<Pa2 K<Pg2 P<Pb3 ...
	R=pa7 R=ra8 q=Pc2 r=Ra1 r=Pa2
static_pawns	Ig2 Ig3 id5 Lc2 Sg2-g3 sg6-g7 P(2) p(3)
dynamic_general	?px ?0x ?p+ ?# ?S !Sr !#b !#n !#q !#r
	!#bn !#bq !#br !#nq !#nr !#qr
dynamic_solution	!-R !-k !-Q !-k !-Q !-Rk !-kQ !-Qk !-kQ
	!xR !xp !xRp
	!k>R !R>k !Q>k !p>Q !Q>r

3.1 Static Features

The static part of the encoding includes information about the positions of pieces, structural relationships between pieces and pawn structures present in the position. The implementation is based on previous work on similar position retrieval [5] and our early work [1] and is intended to serve as a baseline on which we aim to improve by implementing encoding of dynamic features.

Piece Positions and Connectivity. The section describing piece positions and connectivity encoding consists of three parts:

- *naive encoding* - the positions of all the pieces on the board.
- *reachable squares* - all squares reachable by pieces on the board in one move, with decreasing weight based on distance from the original position.
- *connectivity between the pieces* - the structural relationships between the pieces in the positions. For each piece it is recorded which other pieces it attacks, defends or attacks through another piece (*X-ray attack*).

Pawn Structures. For the static part of the encoding, we also use pawn structure detection algorithms to detect the following pawn structures in the position and encode them into terms: isolated pawns, (protected) passed pawns, backward pawns, doubled pawns and pawn chains.

3.2 Dynamic Features

In the dynamic part of the encoding, we focus more on the solution of the tactical problem and try to grasp the motif behind it.

We first encode some general features of the solution and then add more specific terms describing the sequence of moves. The types of pieces that are moved or captured in the move sequence and the interactions between pieces are described in the encodings. We also try to identify piece sacrifices in the solution sequence. We are mainly interested in the motifs that occur in the solution and make the encodings of the features independent of where exactly on the chessboard they occur. For this reason, we do not include any exact positions of the pieces in the textual terms, but only describe the pieces by their types.

General Dynamic Features. In this part we encode some basic features of the solution move sequence that can help us determine similarity. We use a single term for each of the following features if it holds for the solution:

- *?px* - the player captures a piece in at least one of the moves
- *?ox* - the opponent captures a piece in at least one of the moves
- *?+* - the player gives a check at least once during the sequence
- *?=* - the player promotes a pawn in at least one of the moves
- *?S* - the player sacrifices one or more pieces
- *?#* - the solution ends with a checkmate
- *?1/2* - the solution ends in a draw

Solution Sequence Features. In this section we encode information about the solution move sequence. The encoding includes a term for each:

- type of piece moved: *!-{piece symbol}*
- type of piece captured: *!x{piece symbol}*
- attack between pieces that occurs during the solution: *!{attacking piece symbol}>{attacked piece symbol}*
- type of piece sacrificed: *!S{piece symbol}*
- type of piece involved in checkmate: *!#{piece symbol}*

We count a piece as involved in checkmate if it is attacking either the king directly or any of the squares adjacent to the king. To include information about the order of moves and captures we also include a term for each two consecutive moves and captures in the solution. We also include a term for each pair of pieces involved in checkmate to capture more specific combinations of pieces.

4 Building a Database of Tactical Puzzles

To build a system that can recommend relevant tactical problems for any given game, a large enough collection of tactics to draw from is as important as the search algorithm itself. To obtain a larger dataset for our experiments, we developed the following method to automatically traverse a collection of games and generate tactical puzzles from positions with certain properties.

- To find the most relevant candidate positions, we analyse sequences of three positions. We look for the pattern where the first player makes a blunder (the engine's rating changes by more than a certain threshold), which is immediately followed by another blunder by the second player. This means that the correct move in the position is probably not obvious, giving us a potentially interesting tactical problem. The change in evaluation must be significant and effectively change the theoretical game value.
- We analyse candidate positions with a chess engine to find the optimal line of play and other relevant moves in order to determine whether they are suitable as tactical problems and to find solutions to them. We also determine the correct length of a solution, which is another difficult problem.
- Finally, we perform an additional filtering step. Namely, the solution must be the only unambiguously winning move sequence from the initial position, the player must gain a clear material advantage or checkmate the opponent, and the final position must be stable.

5   Evaluation

To evaluate the effectiveness of our approach, we conducted three experiments. In the first experiment, we took a large number of pairs of similar tactical problems from a renowned chess tactics training course. The pairs of similar training examples in this course were determined by chess experts. The task of our program was to find the pairs based on similarity calculations. In the second experiment, our program picked out the most similar chess tactics to the queried chess tactics and a chess expert was asked to explain the reasons for the similarity. In the third experiment, a group of experts was asked to find the most similar chess tactics to the queried chess tactics and the results were compared to the results of the program.

5.1   Matching Pairs of Puzzles from a Chess Training Course

In the first experiment, we use a number of problems that we have collected from the Chess Tactics Art (CT-ART 6.0) training course. Many puzzles in this course consist of pairs of positions: one is taken from a real game, another represents a simplified version where the same tactical motif usually appears on a smaller 5×5 board. This fact allowed us to obtain a set of position pairs that were considered similar by human experts. We manually checked the puzzles and verified the similarity between the solutions of the individual problem pairs. A total of 400 pairs were collected for the test data set.

An example of such a pair is shown in Fig. 2. The solution to both problems is to sacrifice the rook on the e-file to remove the defender of the g7 square, resulting in checkmate with the queen. The solution in the simplified problem contains the same motif, but there are much fewer pieces, so the solution is generally easier for the students to find.

static	30.85
dynamic	53.30
total	**84.15**
r<ne8	10.90
R>ne8	4.62
!xnR	4.35
!#bq	4.03
!-rQ	3.97
...	

Fig. 2. A pair of tactical problems from the data set: base problem (left) and simplified problem (right). The solution is in both cases 1.♖ xe8 followed by 2.♕ g7 checkmate.

We first build an index using the simplified version of the problem from each pair, then perform a query on the index with each of the regular problems. For each query we record the rank of the matching position in the results and calculate how often the matching position appears as the top result or within the first N results.

We tested the search accuracy using the following feature subsets: all static features, all dynamic features and all features combined. All runs use the default BM25 parameters $k_1 = 1.2$ and $b = 0.75$ and all included feature sets are weighted equally. The results are presented in Table 2.

Table 2. Success rates for different configurations.

Feature set	Accuracy		
	top-1	top-5	top-10
all static features	0.252	0.370	0.433
all dynamic features	0.418	0.652	0.761
all features	**0.481**	**0.736**	**0.814**

Using either static or dynamic features only does not yield the best results. The results are significantly improved when both static and dynamic features are combined. This shows that each set of features covers a different aspect of the tactic, both of which need to be considered when determining similarity.

5.2 Chess Expert's Explanations of Similarity

In the second experiment, we selected 10 contextually different chess tactical problems and then automatically retrieved 5 most similar positions for each of them from a database of 46,370 tactical problems constructed from the lichess.org game database.

The resulting most similar positions were shown to a chess expert. The expert was asked to comment on the reasons for the similarity of the resulting problems with the original query positions, taking into account both static and dynamic aspects. The expert was able to explain the similarity in 48 out of 50 problems. Overall, the expert praised the program's ability to detect dynamic similarity of positions, even if the initial positions differ significantly.

Table 3. The results of the experiment with three chess experts.

				Expert 1		Expert 2		Expert 3	
ID	Top score	Avg. score	SD	Score	Rank	Score	Rank	Score	Rank
1	63.42	15.41	11.04	63.42	1	63.42	1	63.42	1
2	82.09	15.60	17.13	81.47	2(1)	81.47	2(1)	81.47	2(1)
3	72.13	22.19	15.56	60.67	2(2)	60.67	2(2)	60.67	2(2)
4	72.32	15.72	13.46	72.32	1	72.32	1	72.32	1
5	91.58	22.55	17.43	89.17	2(1)	89.17	2(1)	89.17	2(1)
6	78.05	16.08	15.20	78.05	1	78.05	1	78.05	1
7	72.67	16.07	13.04	72.67	1	72.67	1	72.67	1
8	69.03	23.22	14.55	69.03	1	69.03	1	69.03	1
9	78.77	24.69	14.54	78.77	1	78.77	1	61.58	2(2)
10	70.37	16.10	13.23	70.37	1	25.72	6(11)	70.37	1

5.3   Comparison with a Group of Chess Experts

In the third experiment, we compared the choice of the most similar tactics according to the program with the choice according to three chess experts: a woman grandmaster, a master, and a strong club player.

We first selected 20 contextually distinct chess tactical problems and then automatically searched for three similar chess tactical problems using (1) static features only, (2) dynamic features only, and (3) all features. Half of these contextually different chess tactics were presented to a group of chess experts. The found chess tactics were presented to the experts in another database. In this database, 7 duplicates occurred because the program retrieved the same chess tactics with dynamic features or with all features. After removing the duplicates, 53 tactics remained for comparison. The experts' task was to look at the original 10 tactics and find the most similar tactic for each of them in the database of 53 tactics retrieved by the program.

The results of the comparison are shown in Table 3. Each row in the table first shows ID of a query tactic and the results according to the program: the highest score of the most similar tactics, the average score of all tactics in terms of similarity to the query item, and the standard deviation of these scores. Then the results of the three experts are displayed: the program's score of the most similar tactic according to a chess expert and the program's rank of this tactic

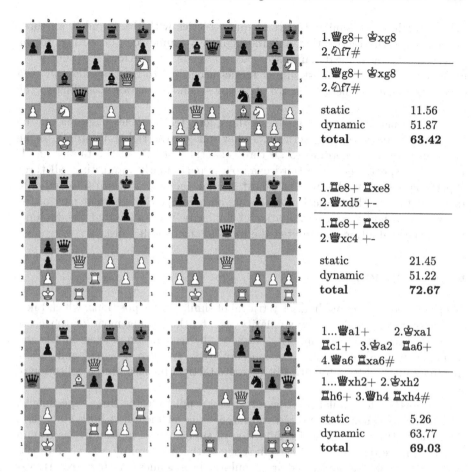

Fig. 3. Similar tactics with corresponding solutions and scores.

among all 53 candidate tactics. Spearman's statistical test showed that there is a significant correlation between the rankings of the program and the rankings of the chess experts ($\rho = .645$, $p = .001$).

In the cases where the experts did not choose the same most similar tactic as the program, we also show the rank of the tactic according to its dynamic score.

The results show that all three experts strongly agreed not only with the program's choices, but also with the choices of the other experts. With only one exception, the experts always chose the most similar or second most similar tactic according to the program. The exception was when the second expert chose a tactic that was similar to the query tactic with ID 10. But even in this case, the expert pointed to the most similar tactic according to the program as an alternative (which also shows that the decision was not easy).

We also found that the dynamic part of the score played the most important role in the evaluation of similarity by both the program and the experts.

When searching for similar tactics to query tactics with IDs 2 and 5, all three experts chose the tactics that had the highest dynamic score according to the program. Moreover, the dynamic score accounted on average for 77% (SD = 0.11) of the total score of the most similar tactics according to the program and even 80% (SD = 0.10) of the total score of the most similar tactics according to the majority vote of the experts.

Figure 3 shows three examples of the query tactics (the diagrams on the left) and the most similar tactics (the diagrams on the right) according to the program and also according to all three experts (the IDs of the tactics are 1, 7, and 8, respectively). The third example is particularly interesting: all three experts agreed that this was the most similar tactic among all 53 candidate tactics, although the tactical motif occurs on a completely different part of the board. From the program's evaluation of similarity, we can see that the dynamic part of the evaluation contributed the most to the overall score.

6   Conclusion

We presented a novel method for retrieval of similar chess positions, which takes into account not only static similarity due to the arrangement of the chess pieces, but also dynamic similarity based on the recognition of chess motifs and dynamic, tactical aspects of position similarity. We also designed and implemented a method to automatically generate tactical puzzles from a collection of games.

The method for similar position retrieval was put to test in three experiments. The first experiment emphasised the importance of including both static and dynamic features for successful detection of similar chess motifs. In the second experiment we demonstrated the efficiency of the program on a large database of tactical problems generated from online chess games. A chess expert was able to explain the similarity in the vast majority of the retrieved problems and praised the program's ability to detect dynamic similarity of positions even if the initial positions differ significantly. The results of the third experiment showed that all three experts are congruent not only with the choices of the program, but also with the choices of each other. Importantly, the experiments clearly demonstrated that the dynamic part of the score played the most important role in the evaluation of similarity by both the program and the experts.

The resulting program can be useful for the automatic generation of instructive examples for chess training. Our approach is certainly not limited to chess.

References

1. Bizjak, M., Guid, M.: Towards automatic recognition of similar chess motifs. In: Proceedings of Slovenian Conference on Artificial Intelligence 2020, pp. 11–14 (2020)
2. Chess Informant: Encyclopedia of Chess Combinations. Chess Informant (2014)
3. Costeff, G.: The chess query language: CQL. ICGA J. **27**(4), 217–225 (2004)
4. Dvoretsky, M., Yusupov, A.: Secrets of Chess Training. Edition Olms (2006)

5. Ganguly, D., Leveling, J., Jones, G.J.: Retrieval of similar chess positions. In: Proceedings of the 37th International ACM SIGIR Conference on Research and Development in Information Retrieval, pp. 687–696. ACM (2014)

6. Guid, M., Možina, M., Krivec, J., Sadikov, A., Bratko, I.: Learning positional features for annotating chess games: a case study. In: van den Herik, H.J., Xu, X., Ma, Z., Winands, M.H.M. (eds.) CG 2008. LNCS, vol. 5131, pp. 192–204. Springer, Heidelberg (2008). https://doi.org/10.1007/978-3-540-87608-3_18

7. Robertson, S.E., Walker, S., Jones, S., Hancock-Beaulieu, M.M., Gatford, M., et al.: Okapi at TREC-3. NIST Special Publication SP **109**, 109 (1995)

8. Woolf, B.P.: Building Intelligent Interactive Tutors: Student-Centered Strategies for Revolutionizing E-learning. Morgan Kaufmann, Burlington (2010)

9. Zloof, M.M.: Query-by-example: the invocation and definition of tables and forms. In: Proceedings of the 1st International Conference on Very Large Data Bases, pp. 1–24 (1975)

On the Road to Perfection? Evaluating Leela Chess Zero Against Endgame Tablebases

Rejwana Haque$^{(\boxtimes)}$, Ting Han Wei, and Martin Müller

University of Alberta, Edmonton, Canada
{rejwana1,tinghan,mmueller}@ualberta.ca

Abstract. Board game research has pursued two distinct but linked objectives: solving games, and strong play using heuristics. In our case study in the game of chess, we analyze how current AlphaZero type architectures learn and play late chess endgames, for which perfect play tablebases are available. We study the open source program Leela Chess Zero in three and four piece chess endgames. We quantify the program's move decision errors for both an intermediate and a strong version, and for both the raw policy network and the full MCTS-based player. We discuss a number of interesting types of errors by using examples, explain how they come about, and present evidence-based conjectures on the types of positions that still cause problems for these impressive engines.

Keywords: AlphaZero learning · Computer chess · Leela Chess Zero

1 Introduction

The AlphaZero algorithm [12] has demonstrated superhuman playing strength in a wide range of games, such as chess, shogi, and Go. Yet as powerful as neural networks (NNs) are at move selection and state evaluation, they are not perfect. Judging from the varied outcome of self-play games in deterministic complete information games such as chess and Go, even the best AlphaZero-style players must still make mistakes. We investigate this gap between strong play and perfect play. To analyze how these modern programs learn to play sophisticated games, and also to test the limits to how well they learn to play, we turn to a sample problem that has known exact solutions. While the full game of chess has not yet been solved, exact solutions for endgames up to seven pieces have been computed and compiled into endgame tablebases. We use the AlphaZero-style open-source program *Leela Chess Zero* (Lc0) to analyze chess endgames. We develop a methodology and perform large-scale experiments to study and answer the following research questions:

- How do stronger and weaker networks differ for predicting perfect play?
- How does the search in Lc0 improve the prediction accuracy for endgames?
- How do stronger policies improve search results?

© Springer Nature Switzerland AG 2022
C. Browne et al. (Eds.): ACG 2021, LNCS 13262, pp. 142–152, 2022.
https://doi.org/10.1007/978-3-031-11488-5_13

- How does reducing the search budget affect the correctness of Lc0?
- Which is easier to predict, wins or draws?
- How well does Lc0 recognize wins and losses?
- Which kinds of endgame positions are easy and hard to learn? How does that change with more learning?
- Are there cases where search negatively impacts move prediction? If such cases exist, why do they occur?

2 Background

In previous work [6,11], perfect play has been compared against heuristic engines.

2.1 Chess Endgame Tablebases

Chess is a game in which the state space and game tree complexity is reduced as the game progresses and pieces are captured. Chess endgames are sub-problems in which the full rules of chess apply, but only a reduced set of game pieces remains on the board. While the game of chess itself has not been solved to date, endgames of up to seven pieces have been solved and are publicly available [9]. A database of such endgame solutions is referred to as an endgame tablebase. A solution for each position includes the outcome of the game given perfect play for both players, the optimal moves that each player must make to reach that outcome, and specific metrics such as the number of plies (moves by one player) required to reach the outcome. There are advantages and disadvantages to each kind of metric. In this paper, we use the *depth to mate* (DTM) metric, which is the number of plies for a win or loss, assuming the winning side plays the shortest way to win, and the losing side the longest to lose [4].

Endgame tablebases hosted online [5,10] differ in storage size and metrics. Tablebase generators are often also available. Among these, the Syzygy [10] and Gaviota [1] tablebases are free and widely used.

2.2 Leela Chess Zero

Leela Chess Zero (Lc0) is an adaptation of the Go program Leela Zero for chess. Both programs are distributed efforts which reproduce AlphaZero for chess and Go, respectively. Volunteers donate computing resources to generate self-play games and optimize the NNs. Relying solely on community resources and efforts, in 2020 Lc0 surpassed AlphaZero's published playing strength in chess [7].

Like AlphaZero, Lc0 also takes a sequence of consecutive raw board positions as input. It uses the same two-headed (policy and value) network architecture as AlphaZero. Similar to AlphaZero, the policy head output guides the Monte Carlo tree search (MCTS) as a *prior probability*, while the value head output replaces rollouts for *position evaluation*. Over time the developers of Lc0 introduced enhancements that were not in the original AlphaZero. For example, an auxiliary output called the *moves left head* was added to predict the number of plies remaining in the current game [3]. Another auxiliary output called the *WDL* head separately predicts the probabilities that the outcome of the game is a win, draw, or loss [8]. Lc0 uses two distinct training methods to generate different types of networks that differ in playing strength. T networks are trained by self-play, as in AlphaZero, while J networks are trained using self-play game records generated from T networks. J networks are stronger and are used in tournament play.

3   Analyzing Chess Endgames with Leela Zero Chess

3.1   Tablebase Dataset Pre-processing

In order to evaluate the performance of a chess program in terms of finding exact solutions, a ground truth to compare against is needed. For that, we use chess endgame tablebases which describe perfect play of all positions up to 7 pieces on the board. In this section we describe how to pre-process the tablebases.

We choose the open-source Gaviota tablebases as the source of ground truth. We compare the perfect play against the tested program's choice for all unique legal positions in all nontrivial three and four piece endgames. Also, we took the winning and drawing positions where there are more than one possible outcome.

Since the Gaviota tablebase is indexed and compressed, the following steps are performed to create an iterable list of positions:

1. We use the method of Kryukov [5] to enumerate all unique legal positions and store positions in the FEN format. Uniqueness refers to treating symmetric positions as one; legality is checked using normal chess rules.
2. We use tools from python-chess [2] to extract the following information from the Gaviota tablebase for each enumerated position: The position in FEN format, lists of all winning, drawing and losing moves, the win-draw-loss status, the DTM score, and the decision depth (see below).
3. The data for each endgame type is stored in MySQL for easy access.

We use the term *decision depth* to categorize positions in an endgame tablebase. For a winning position, the decision depth is simply the DTM score. For drawing positions where there are also losing moves, the decision depth is the highest DTM after a losing move.

3.2   Choice of AlphaZero Program and Its Parameters

We chose Lc0 0.27 as our AlphaZero-style program in our analysis, because it is both publicly available and strong. Lc0 has evolved from the original AlphaZero in many ways, but with the proper configuration, it can still perform similarly. We use two specific settings and leave everything else in the default configuration.

Disabling Smart Pruning: AlphaZero selects the node with the highest PUCB value during MCTS [13]. However, Lc0 does not always follow this behaviour. This is due to smart pruning, which uses the simulation budget differently: it stops considering less promising moves earlier, resulting in less exploration. We set the LcO parameter `-smart-pruning-factor=0` to disable smart pruning.

Single-Threaded Search: For consistent analyses, we prefer the engine to be deterministic between different experiment runs. Multi-threading introduces random behaviour, so we run the engine with a single thread for consistency.

LcO supports a variety of NN backends. Since we used Nvidia Titan RTX GPUs for our experiments, we chose the `cudnn` backend.

Many network instances are publicly available. For this research we chose two specific snapshots of the T60 network generation:

Strong Network: ID 608927 with (self-play) ELO rating 3062.00, which was the best performing snapshot up to May 2, 2021.

Weak Network: ID 600060, rating 1717.00, were the initial weights of this generation after 60 updates starting from random play.

Table 1. Total number of mistakes by the policy net and MCTS with 400 simulations, using strong and weak networks.

EGTB	Total Positions Tested	Weak Network		Strong Network	
		Policy	MCTS-400	Policy	MCTS-400
KPk	8596	390	13	5	0
KQk	20743	109	0	0	0
KRk	24692	69	0	0	0
KQkq	2055004	175623	12740	3075	36
KQkr	1579833	141104	3750	4011	46
KRkr	2429734	177263	6097	252	0
KPkp	4733080	474896	41763	20884	423
KPkq	4320585	449807	46981	6132	13
KPkr	5514997	643187	60605	13227	196

4   Experiments

4.1   Move Prediction Accuracy for Basic Settings

We evaluate the move decisions of LcO for all non-trivial three and four piece endgame positions. We define a *mistake* as a move decision that changes the game-theoretic outcome, either from a draw to a loss, or from a win to not a win. Any outcome-preserving decision is considered correct. So the engine does not need to choose the quickest winning move. Accuracy is measured in terms of the number or frequency of mistakes, over a whole database.

Table 1 shows the number of mistakes, for each three and four piece tablebase, for both the weak and the strong network. Results are shown for both the raw network policy, and for full Lc0 with 400 MCTS simulations. From the table it is clear that both network training and search work very well. The strong net makes significantly fewer mistakes, and search strongly improves performance in each case where the network alone is insufficient. For all three piece positions, 400 MCTS simulations are enough to achieve perfect play. In four piece endgames, a small number of mistakes still remain under our test conditions.

Effect of Search Budget on Winning and Drawing Positions. To analyze how deeper search influences accuracy, we compare search budgets of 0 (raw policy), 400, 800, and 1600 simulations per move decision. We call these settings MCTS-0 = policy, MCTS-400, ... We chose these relatively settings considering our limited computational resources. Deeper search consistently helps for all of these tablebases.

The error rate, defined as the fraction of mistakes in a set of positions, is shown separately for the sets of winning and drawing positions in each tablebase in Table 2. In many of these datasets, decisions are more accurate for draws than for wins. The main exception is KQkr, which contains only a small fraction (3.4%) of draws. The error rate for those draws is very high at 2%.

Table 2. Error rate for winning and drawing move predictions for MCTS with search budgets of 0 (raw policy), 400, 800 and 1600 simulations.

EGTB	% of Error							
	Winning Positions				Drawing Positions			
	0	400	800	1600	0	400	800	1600
KPk	$8.00e-2$	0.00	0.00	0.00	0.00	0.00	0.00	0.00
KQk	0.00	0.00	0.00	0.00	0.00	0.00	0.00	0.00
KRk	0.00	0.00	0.00	0.00	0.00	0.00	0.00	0.00
KQkq	$2.96e-1$	$3.21e-3$	$1.61e-3$	0.00	$2.74e-2$	$5.35e-4$	$8.03e-4$	$5.35e-4$
KQkr	$1.53e-1$	$1.41e-3$	$4.01e-4$	0.00	2.01	$2.93e-2$	$1.41e-2$	$1.17e-3$
KRkr	$2.33e-2$	0.00	0.00	0.00	$4.19e-3$	0.00	0.00	0.00
KPkp	$4.39e-1$	$1.02e-2$	$3.30e-3$	$1.46e-3$	$4.45e-1$	$6.18e-3$	$1.51e-3$	$3.29e-4$
KPkq	$1.18e-3$	$2.94e-6$	$1.34e-6$	$10.07e-6$	$2.94e-3$	$3.42e-6$	0.00	0.00
KPkr	$2.02e-1$	$4.11e-1$	$3.25e-3$	$4.92e-3$	$1.22e-3$	$1.81e-3$	$2.88e-4$	$1.31e-3$

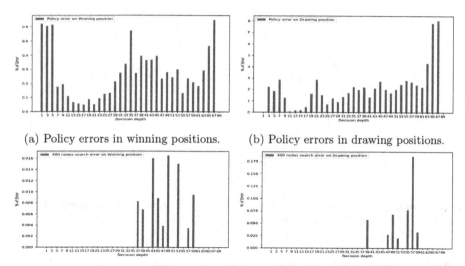

(a) Policy errors in winning positions.   (b) Policy errors in drawing positions.

(c) MCTS-400 errors in winning positions. (d) MCTS-400 errors in drawing positions.

Fig. 1. Error rate for each decision depth in the KQkr tablebase.

4.2   Performance at Different Decision Depths

In this experiment, shown in Fig. 1, we measure the error rate separately at each decision depth. We evaluate both the raw policy and MCTS-400 using the strong network. The figure shows that in contrast to raw policy, MCTS-400 only makes mistakes at higher decision depths. Policy mistakes at all shallow decision depths are completely corrected by search. At higher depths, some errors remain, but there is no simple relation between decision depth and error rate there.

Figure 2 shows that there is a relationship between the sample size at each decision depth and the error rate of the raw net. Each point in the figure corresponds to all positions of a specific decision depth in KQkr. The results in other four piece tablebases are similar in that fewer positions at a given depth correspond to more errors. They are omitted here for brevity. Figure 1(c)–(d) shows the corresponding results for MCTS-400. The engine only makes mistakes in positions with higher decision depths. Search can routinely solve positions with shallower decision depths regardless of the policy accuracy.

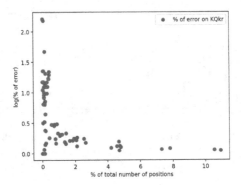

Fig. 2. Sample size (shown as percentage of the total number of positions) at each decision depth vs. raw policy error rate (in log scale) for the KQkr tablebase.

4.3   Case Studies: Interesting Engine Mistakes

In this section we study a number of interesting cases where Lc0 makes mistakes. While analyzing these mistakes, there is a large amount of common expected behavior: search typically corrects policy inaccuracies, and larger searches correct errors that are still made by smaller searches. However, there are cases where search fails while the policy is correct. Several examples discussed below are shown in Fig. 3. The correct moves are indicated in green and blue, while chosen incorrect moves are shown in red.

Policy Wrong, Search Correct: In Fig. 3(a), **Qg1** wins but **Qa1** only draws. The network's prior probability (policy head) of the incorrect move **Qa1** (0.1065) is higher than for the winning move **Qg1** (0.0974). However, the value head has a better evaluation for the position after the winning move (0.3477) than the drawing move (0.0067). Therefore **Qg1** becomes the best-evaluated move after only a few simulations. Figure 4(a1–a4) shows details - the changes of Q, U, $Q + U$ and N during MCTS as a function of the number of simulations. At each simulation, the move with the highest UCB value $(Q + U)$ is selected for evaluation. The Q value of a node is the average of its descendants' values. The exploration term U depends on the node's visit count N and the node's prior probability. For this example, while the exploration term $U(\mathbf{Qa1}) > U(\mathbf{Qg1})$ throughout, the UCB value remains in favour of the winning move. An accurate value head can overcome an inaccurate policy in the search.

Policy and Search Both Wrong: In Fig. 3(b), both **Kd3** and **Kd5** win, but both the raw network and the search choose **Kc3** which draws. **Kd5** has by far the lowest policy (0.2776) and value (0.3855), and its Q and N are consistently low, keeping it in distant third place throughout. Both the initial policy and value are higher for **Kc3** (0.3229 and 0.9144) than for the correct **Kd3** (0.2838 and 0.8501). We extended the search beyond the usual 1600 simulations to see longer-term behavior. The Q value of **Kc3** remains highest for 6000 simulations, while **Kd3** catches up, as shown in Fig. 4(b1). MCTS samples all three moves with

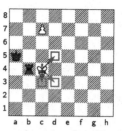

(a) Policy wrong, search correct

(b) Policy wrong, search also wrong

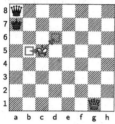

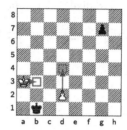

(c) Policy correct, smaller search wrong

(d) Policy correct, all search wrong

Fig. 3. Examples for different types of engine mistakes.

similar UCB values, but focuses most on the incorrect **Kc3**. At 1600 simulations, the inaccurate value estimates play a large role in incorrectly choosing **Kc3**. Beyond 6000 simulations, the Q value of **Kd3** keeps improving, and MCTS finally chooses a correct move at about 12000 simulations.

Policy Correct, Smaller Search Wrong: In Fig. 3(c), **Kb5** wins while **Kd6** draws. The prior probability of **Kb5** is 0.0728, which is slightly higher than **Kd6**'s at 0.0702, but the value, at 0.2707, is slightly lower than **Kd6**'s at 0.2834. Figure 4(c1) shows that the Q value of **Kd6** is higher early on due to the value head. As search proceeds, this reverses since the values in the subtree of the winning move are better. In this example, MCTS overcomes an inaccurate root value since the evaluations of its followup positions are more accurate.

Policy Correct, Search up to 1600 Simulations Wrong: In the example shown in Fig. 3(d), **d4** wins. Up to 1600 simulations, MCTS chooses the drawing move **Kb3**. The value of **Kb3** (0.1457) is higher than that of **d4** (0.1053), but the prior probability of **d4** (0.3563) is higher than **Kb3** (0.348). Figure 4(d1–d4) shows the search progress. The Q value of **d4** remains lower for longer than in the previous example in Fig. 3(c). At around 1500 simulations, the UCB value of the correct move becomes consistently higher. This prompts the search to sample the correct move more, At 2200 simulations, the Q value of the correct **d4** spikes dramatically. At this point the search tree is deep enough to confirm the win, and from 2700 simulations on, the engine plays **d4**.

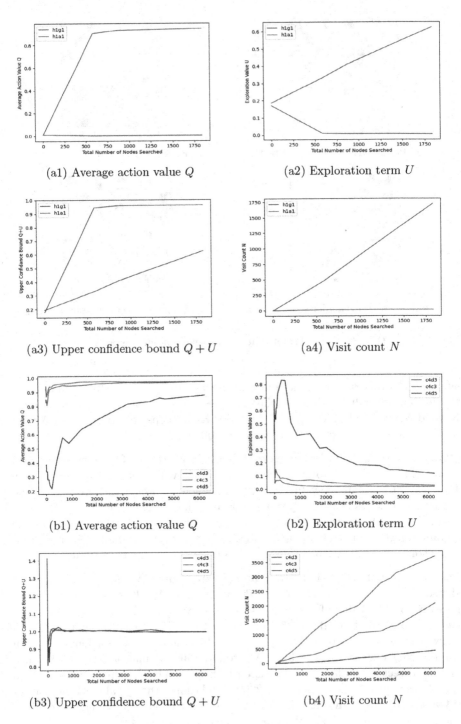

(a1) Average action value Q

(a2) Exploration term U

(a3) Upper confidence bound $Q + U$

(a4) Visit count N

(b1) Average action value Q

(b2) Exploration term U

(b3) Upper confidence bound $Q + U$

(b4) Visit count N

Fig. 4. Development of relevant terms Q, U, $Q + U$, N in UCB for Fig. 3((a)–(d)).

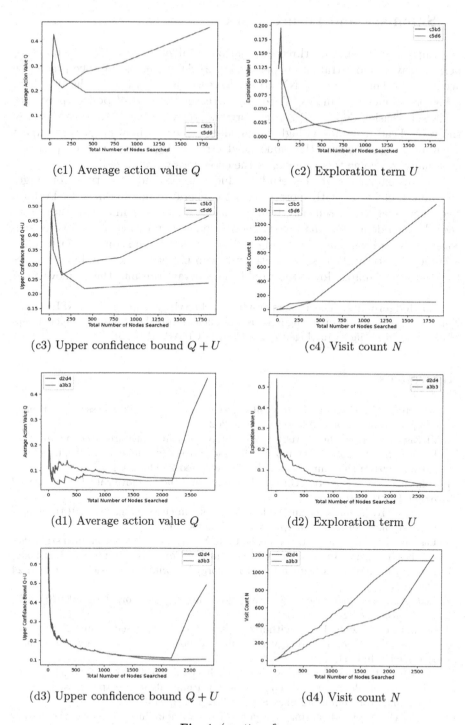

(c1) Average action value Q

(c2) Exploration term U

(c3) Upper confidence bound $Q + U$

(c4) Visit count N

(d1) Average action value Q

(d2) Exploration term U

(d3) Upper confidence bound $Q + U$

(d4) Visit count N

Fig. 4. (*continued*)

5   Summary and Future Work

The important findings for three and four piece tablebases are: 1) NNs approach perfect play as more training is performed. 2) Search helps improve prediction accuracy. 3) The number of NN errors decreases for decision depths that have a higher number of samples. 4) Search increases the rate of perfect play with shallower decision depths. 5) Search corrects policy inaccuracies in cases where the value head accuracy is high. 6) Small-scale search may negatively impact accuracy in cases where the value head error is high. However, deep search eventually overcome this problem in the endgames we analyzed.

Future extensions of this study include: 1) Extend the study for larger endgame tablebases (with more pieces) to generalize our findings. 2) Perform frequency analyses of self-play training data to get the number of samples at each decision depth. 3) Analyze symmetric endgame positions to verify decision consistency. 4) Examine value head prediction accuracy and compare with policy accuracy. 5) Study the case where the program preserves the win, but increases the distance to mate. How often would the program run into the 50 move rule?

Acknowledgements. The authors gratefully acknowledge support from NSERC, the Natural Sciences and Engineering Research Council of Canada, and from Müllers Deep-Mind Chair in Artificial Intelligence and Canada CIFAR AI Chair.

References

1. Ballicora, M.: Gaviota. https://sites.google.com/site/gaviotachessengine/Home/endgame-tablebases-1. Accessed 6 June 2021
2. Fiekas, N.: python-chess, August 2014. https://python-chess.readthedocs.io
3. Forstén, H.: Purpose of the moves left head. https://github.com/LeelaChessZero/lc0/pull/961#issuecomment-587112109. Accessed 24 Oct 2021
4. Huntington, G., Haworth, G.: Depth to mate and the 50-move rule. ICGA Journal **38**(2), 93–98 (2015)
5. Kryukov, K.: Number of unique legal positions in chess endgames (2014). http://kirill-kryukov.com/chess/nulp/. Accessed 30 Oct 2021
6. Lassabe, N., Sanchez, S., Luga, H., Duthen, Y.: Genetically programmed strategies for chess endgame. In: Genetic and Evolutionary Computation, pp. 831–838 (2006)
7. Lc0 Authors: What is Lc0?. http://lczero.org/dev/wiki/what-is-lc0/. Accessed 24 Oct 2021
8. Lc0 Authors: Win-draw-loss evaluation. http://lczero.org/blog/2020/04/wdl-head/. Accessed 24 Oct 2021
9. de Man, R.: Syzygy. http://github.com/syzygy1/tb. Accessed 6 June 2021
10. de Man, R., Guo, B.: Syzygy endgame tablebases. http://syzygy-tables.info/. Accessed 6 June 2021
11. Romein, J.W., Bal, H.E.: Awari is solved. ICGA J. **25**(3), 162–165 (2002)
12. Silver, D., et al.: A general reinforcement learning algorithm that masters chess, shogi, and go through self-play. Science **362**(6419), 1140–1144 (2018)
13. Silver, D., et al.: Mastering the game of Go without human knowledge. Nature **550**(7676), 354–359 (2017)

Chess Endgame Compression via Logic Minimization

Dave Gomboc[✉] and Christian R. Shelton

University of California, Riverside, CA 92521, USA
dave_gomboc@acm.org, cshelton@cs.ucr.edu

Abstract. Chess endgame tables encode perfect information to inform heuristic search and permit error-free play once the root position is within them. We introduce a new approach to their minimization, and demonstrate better probe performance than the state-of-the-art.

Keywords: Chess · endgame tables · logic minimization

1 Introduction

Chess play has been studied for over a century, decades before computing science itself existed as a discipline [5,8,24,25,27,31,38,41]. Today, machines are substantially stronger Chess players than top human experts, and the same can be said regarding many other similar traditional human games.

1.1 Game-Theoretic Error-Free Endgame Play

A Chess endgame table (EGT) is a precomputed, known-correct source of information about Chess endgame positions. The first Chess EGTs were computed by Ströhlein [36]; seven-piece EGTs were first computed by Zakharov et al. [40] on the Lomonosov supercomputer, using tens of tebibytes. Chess engines that reach such pre-tabulated positions from within their heuristic searches can propagate back an exact score, which can be used either to improve on-line game play or to improve the accuracy and efficiency of off-line reinforcement learning [32] via endgame table rescoring [21].

1.2 State-of-the-Art Chess Endgame Data Compression

Syzygy [22] EGTs are predominantly used at present because they are more compact than any widely-available alternative, while also being acceptably efficient to query. Syzygy has coverage for positions where en passant is possible, but has no data for where some castling right exists. By design, Syzygy stores misleading values for positions containing legal captures that achieve better compression: correct querying of them requires the concomitant use of a capture-based quiescence search and minimaxing the resulting values. We use (an updated version of) Fathom [13] to provide these capabilities.

C. Browne et al. (Eds.): ACG 2021, LNCS 13262, pp. 153–162, 2022.
https://doi.org/10.1007/978-3-031-11488-5_14

For each material balance (the set of pieces remaining, for each player) covered, Syzygy includes both a win-draw-loss (WDL) table, and a distance-to-zeroing-move (DTZ) table. WDL alone is sufficient to avoid entering a game-theoretically-suboptimal position and to determine the result of a game, as has been previously achieved in Checkers [30], so we do not consider DTZ further.

1.3 Approach

A Chess EGT may be viewed as a partial function that maps a subset of Chess positions to game-theoretic outcomes. We employ two-level logic minimization to represent this function in a compact form. Binary decision diagrams [7] have been used for a related purpose in domains such as Connect Four [12]. In Sect. 2, we review two-level logic minimization, and explain how we encode Chess positions as logic bits. We discuss the experiments we undertake in Sect. 3, and summarize our contributions in Sect. 4.

2 Two-Level Logic Minimization

Consider a partial function $P : \{0,1\}^n \rightarrow \{0,1\}^m$. An equivalent total function $T : \{0,1\}^n \rightarrow \{0,1,X\}^m$ exists, where an output of X indicates that we do not care which truth value is assigned to that output. The straightforward tabular representation of T would always contain 2^n rows. For succinctness, we use a matrix representation $M : \{0,1,X\}^n \rightarrow \{0,1,X\}^m$ of P, where an input of X indicates that the row is applicable regardless of the instantiated truth value of that input. Thus, a single row of matrix M with k inputs set to X is equivalent to specifying the 2^k compatible rows of the tabular representation of T.

The union of the input vectors where any of the outputs is assigned to either 1, 0, or X is considered to be part of the ON-cover (or F, for function), the OFF-cover (or R, for reverse), or the DC-cover (or D, for "don't care"), respectively. Each such cover is the sum of clauses; each clause (or "cube") is the product of individual inputs.

Definition 1. Two-level logic minimization *is the task of, having been provided with some matrix M that is consistent with P, identifying a matrix M' that is also consistent with P whose covers of interest have minimum cardinality.*

We first discuss a few important algorithms from the electronic design automation (EDA) literature; see Coudert [9] for coverage of additional historically-important techniques. We then describe the mapping from Chess endgame table data to $\{0,1,X\}$-vectors and how to employ logic minimization in this context.

2.1 MINI

The MINI logic minimizer [20] introduced the heuristic approach of iteratively improving cover cardinality via repeated cube expansion and reduction.

Positional Cube Notation. As with one-hot encodings used in machine learning, positional cube notation (PCN) maps each specific value of an input variable to a different bit. Doing so permits efficient cube operations to be performed via bitwise logical operators. A multiple-valued input variable over the domain {ant, bee, cat} could be mapped as: ant → 100; bee → 010; cat → 001. In contrast to a one-hot encoding, 111 is also valid PCN, representing "do not care". For each binary input variable v, PCN reduces to a bit pair $\bar{v}v$: 0 → 10; 1 → 01; X → 11.

Distance-One Merging. Hong *et al.* [20] report using merging two cubes that disagree on only a single variable for "computational advantage", as in these three examples:

before	0X01 100	100X 010	X011 001
	0X11 100	10XX 010	X0X1 001
after	0XX1 100	10XX 010	X0X1 001

MINI iterates over each such input variable once, updating the sorted ordering of M' prior to processing each variable to ensure that the clauses are ordered to permit all potential merges involving that variable via a linear scan through the product clauses.

Expansion. Distance-one merging is a particular form of cube expansion, which is the process of enlarging a cube so that it (hopefully) includes as many as possible of the minimum product terms, or *minterms*, of M' that must be covered, while avoiding covering any product terms that must not be covered (the collection of which constitutes the *blocking cover*). Any cube that is completely covered by another may be discarded, thereby reducing the cardinality of M'.

Reduction. Once expansion has occurred, many cubes that partially overlap may cover the same minterms. Cube reduction is the process of shrinking a cube while ensuring that it continues to cover all minterms not already covered by any other cube. This can allow the cube to later grow in a different direction for better overall minimization.

2.2 ESPRESSO

Irredundancy (Introduced by Brayton *et al.* [6]). While expansion alone can eliminate many cubes, it does not eliminate any cube that does not end up completely encompassed by a single other cube. The irredundancy pass within ESPRESSO's expansion-irredundancy-reduction main loop exists to prioritize the cardinality minimization of M' via the detection and removal of such cubes that are nonetheless redundant with respect to multiple other cubes in advance of performing any reduction that could cause an available opportunity for cube removal to be forfeited.

Distance-One Merging. Like MINI, the ESPRESSO implementation used does support the ability to apply distance-one merging across multiple variables of the ON-cover in sequence. Though this capability is not enabled by default, the espresso(1) manual page suggests its use.

2.3 Pupik

Pupik [15, 16] is based on processing ternary trees that represent Boolean functions [14]. It repeatedly performs single-variable absorption and complementation to combine cubes. We observe that a single distance-one merge encompasses both of those operations, and that the full procedure described in Fišer *et al.* [15] is *precisely equivalent* to performing distance-one merging over F. The asymptotic analysis performed therein naturally does not account for practical benefits of the matrix representation such as its memory locality and amenability to bit vector operations.

2.4 A Simple Position Encoding Scheme

The Chess position encoding we use retains the traditional top-level division of Chess endgame positions by their material balance to permit straightforward comparison with the substantial body of prior work. More importantly, the scheme selected is relatively *uninformed* about Chess. Not only do our input vectors contain no machine-learned features, they also fail to manually capture basic Chess notions such as whether the player to move is in check or has at least one legal move that can be played. We have stayed far away from using any sort of bitboard representation [1] that could cause logic minimization-based image processing techniques [4,11,29] to become applicable. No counters are used; even the specific material balance in use is not encoded. Furthermore, we make no application of concepts that might assist logic minimization itself such as multiple-valued variables or reflected binary (a.k.a. Gray) coding. By doing so, we hope to convince the reader of the generality of the compression technique.

The input to the logic minimizer contains a matrix description of the universe of discourse: 0 000010 000000 010000 001001 010 is an example row of T. Of the 25 input bits, the first is 1 iff Black is to move. Chess boards contain 2^6 potential piece locations, so a sextet is used to specify the placement of each piece. Pieces are recorded in ♔♕♖♗♘♙♚♛♜♝♞♟ order. The final triplet is a multi-valued variable indicating whether White wins, draws, or loses with best play, or that we do not care. Were we processing the ♔♘♚♟ table, the row above would be interpreted as follows: White is to move; the white king is on c8; the white knight is on a8; the black king is on a6; the black pawn is on b7; White has a draw with best play. The complete table T for each four-piece material balance contains 2^{25} rows.

Note the austere simplicity of this representation. Extensive efforts have been made to identify indexing schemes that include all legal positions for a material balance, but as few additional illegal positions as possible [18,19,23,37]. It is also common for multiple indexing order permutations to be attempted for each material balance: once it is determined which variant turns out to yield the smallest file size after a subsequent layer of block compression is applied, the necessary data required to select which scheme is to be used for decompression is recorded near the beginning of the file. Instead, we rely upon logic minimization to combine adjacent cubes with compatible outputs.

This representation also permits labelling large blocks of positions with the same output vector *a priori*. For example, all positions where a black pawn is on the eighth rank are illegal. We could specify that we do not care about any such positions within the ♔♘♚♟ table using a single matrix row: X XXXXXX XXXXXX XXXXXX 000XXX XXX. For simplicity, we do not currently represent either castling or en passant rights.

2.5 Method

We construct one matrix M per material balance, where each row represents a (possibly illegal) position and its associated game-theoretic outcome. We then employ logic minimization to construct a more compact matrix M'. We begin with iterated distance-one merging. Then, one or more ESPRESSO operators (e.g., expansion) are applied.

To probe the game-theoretic value of a position, the query position is encoded as aforementioned. Then, the M' for the appropriate material balance is scanned linearly until a match is found. The output bits of the matching entry dictate the returned result.

3 Experimentation

We explore the trade-off between the minimization time and quality, then compare the resulting on-disk size and time to query endgame positions our method and Fathom.

3.1 Two- and Three-Piece Endgame Tables

Our first experiment manipulates three processing conditions while processing the two- and three-piece tables: whether iterated distance-one merging is or is not performed; whether a single expansion pass or the full ESPRESSO algorithm will be executed; whether or not canonicalization is used. This last condition is explained immediately below, followed by discussing the results of this first experiment.

Canonicalization. Symmetries in Chess endgames (and in other puzzles and games) have long been exploited [2,10,17,26,28,33–35]. A simply example of symmetry exploitation is that the Syzygy (and our) EGTs do not include the ♔♚♕ material balance. When needed, the ♔♕♚ table is probed using the inverted position, and the result translated. Additional symmetries do exist, especially in pawnless endgames.

For each equivalence class of positions defined by the available symmetries for a material balance, we designate one in particular as the canonical representation for which WDL data is recorded. All other positions within the equivalence class are assigned exclusively to the DC-cover. The probing operation must then translate to the canonical position during querying.

Results. We perform experiments on two- and three-piece endgames. Regardless of the processing condition under test, for each of the trivially-drawn material balances (♔♚, ♔♗♚, and ♔♘♚), the ON-cover consistently resolves to a single row with all inputs marked as don't care and the outputs indicating a drawn result. We are nonetheless able to observe striking performance differences with the remaining material balances (♔♕♚, ♔♖♚, and ♔♙♚).

Cumulatively, 194 602 clauses in F exist between the six endgame tables under consideration prior to iterated distance-one merging; versus just 29 920 clauses in F afterward. This early 6.5× reduction in cubes yields a significant processing time advantage for the immediately following expansions, and moreover, takes negligible time to perform. Accordingly, we always apply iterated distance-one merging in our experiments hereafter, and not just to F, but also to R and to D.

Canonicalization causes a large majority of positions in each endgame under consideration to be classified as don't care, which substantially increases the opportunities for cube expansion, and thus for ON- and OFF-cover cardinality minimization. We do acknowledge that the use of canonicalization effectively slips some slight amount of domain knowledge into our lossless compression algorithm. However, symmetry exploitation is not inherently Chess-specific, and this same (and even more) domain knowledge is also exploited by Syzygy EGTs, which serves as our standard for comparison. As with iterated distance-one merging, the benefit is so clear that we always apply this process hereafter.

Executing the complete ESPRESSO algorithm improves ON-cover cardinality versus performing only a single expansion pass, but the associated time penalty is noticeable. Thus, we proceeded to further investigate this tradeoff with larger material balances.

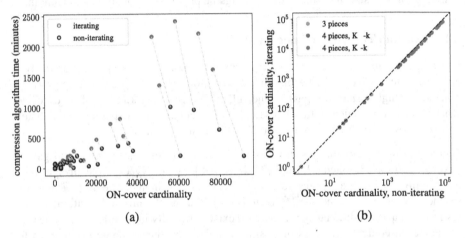

(a)   (b)

Fig. 1. (a) Time versus compression comparison for running ESPRESSO to completion versus performing just a single expansion and irredundancy pass. Each data pair, which is denoted by the dashed line connecting two dots, represents an endgame. The input to all endgames depicted was preprocessed using both iterated distance-one merging and canonicalization. (b) A comparison of the resulting cardinality of the ON-cover for the iterating versus the non-iterating treatment, using the same data as (a). Four-piece endings with two pieces per side tend to be more complex than those where one side has only a king, and so tend to require more cubes to be accurately represented.

3.2 Two- Through Four-Piece Results

We now additionally include four-piece endgames in our exploration. The processing treatment that has been added is to apply iterated distance-one merging, expansion, and irredundancy, but without subsequent iteration of the ESPRESSO reduce-expand-irredundant main loop.

We observe that applying a single irredundancy pass after the expansion pass is both quick and effective at reducing the cardinality of F. Figure 1(a) illustrates that iterating the main loop of ESPRESSO extracts improvements only at a relatively high cost in time; Fig. 1(b) provides an alternate view showing clearly that the ON-cover cardinality reduction achieved for this extra effort is not great. Furthermore, there is every reason to expect that iteration time will increase with problem size.

Exerting additional effort to continue to reduce cover cardinality may be particularly valuable in the electronics manufacturing context. For example, simpler circuits are associated with using either fewer lookup tables (LUTs), or less die space and power. However, the extra effort of iterating until improvement can no longer be found does not appear to be an efficient use of our limited processing power. Given our intention to obtain usable compressed PCN versions of larger EGTs as economically as possible, applying iterated distance-one merging, expansion, then irredundancy appears to be our best trade-off.

3.3 Compression Effectiveness

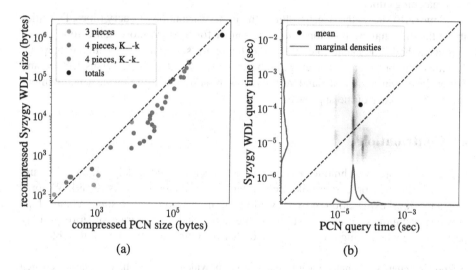

Fig. 2. Comparison of our method versus Syzygy WDL. (a) Table sizes: Each point represents one material balance; points above the diagonal line represent tables for which our method outperforms Syzygy. Note the log-scale: tables to the right and top dominate the total compression size. (b) Query performance: Joint density (across 10M random queries) of the query time for our method and Syzygy. Again, note the log-scales. Our method has a faster mean query time (39 μs versus 138 μs) and smaller standard deviation (41 μs, compared with 258 μs).

Each product clause in F generated when using iterated distance-one merging, expansion, and irredundancy is representable in PCN in 64 bits with room to spare. We generate a binary file per material balance containing each row in PCN sorted lexicographically, then compress it via xz -9 -e. Figure 2(a) contrasts the space consumed on disk by our method with the Syzygy endgame tables after being recompressed with xz -9 -e. For the 2-, 3-, and 4-piece endings, our method does require 47% more space than the recompressed Syzygy EGTs. Given that there has been a half-century of endgame table technology development leading to the Syzygy format, and that numerous opportunities to improve compression results using our novel method remain, we are comfortable claiming that this method of lossless compression has some promise.

3.4 Query Effectiveness

We sampled uniformly with replacement to obtain ten million four-piece Chess endgame positions. We place a white king on any of 64 squares, then a black king on any of 64 squares, then any non-king on any of 64 squares, then any non-king on any of 64 squares, then select the side to move. Any non-legal position is then discarded. Note that castling and en passant rights are never present.

All ten million positions are first probed to verify correctness of the result returned. Afterwards, all ten million positions are probed a second time for timings capture. Fathom is used to probe Syzygy; our method linearly scans the minimized ON-cover for a matching cube.

Figure 2(b) shows the density plot of the joint distribution of query times. Densities above the diagonal are positions that took longer for Syzygy; those below took longer for our method. The mean time (above the diagonal) is shown, along with the marginal densities. Our probes are on average both substantially faster (39 μs versus 138 μs) and exhibit lower variability (a standard deviation of 41 μs versus 258 μs), which demonstrates the viability of the approach.

4   Contributions

Logic minimization techniques have previously been applied widely within EDA, and also within image processing-like and stream compression contexts [3,39]. We provide a top-level explanation of two-level logic minimization, clarify some relationships between techniques described within its literature, and demonstrate superior probing performance when using this method to losslessly compress Chess endgame tables.

Acknowledgments. Computational resources were provided through Office of Naval Research grant N00014-18-1-2252. The first author is also an employee of Google LLC, however, this research has been performed with neither awareness of any applicable insider Google LLC knowledge that might exist nor use of resources associated with that employment. Any statements and opinions expressed do not necessarily reflect the position or the policy of either Google LLC or the University of California; no official endorsement should be inferred.

References

1. Adel'son-Vel'skii, G.M., Arlazarov, V.L., Bitman, A.R., Zhivotovskii, A.A., Uskov, A.V.: Programming a computer to play chess. Russ. Math. Surv. **25**(2), 221–262 (1970). https://doi.org/10.1070/rm1970v025n02abeh003792

2. Allis, L., van der Meulen, M., van den Herik, H.: Databases in awari. In: Levy, D., Beal, D. (eds.) Heuristic Programming in Artificial Intelligence 2: The Second Computer Olympiad, Ellis Horwood, pp. 73–86 (1991)

3. Amarú, L., Gaillardon, P., Burg, A., De Micheli, G.: Data compression via logic synthesis. In: Nineteenth Asia and South Pacific Design Automation Conference (ASP-DAC), pp. 628–633 (2014)

4. Augustine, J., Feng, W., Makur, A., Jacob, J.: Switching theoretic approach to image compression. Sig. Process. **44**(2), 243–246 (1995). https://doi.org/10.1016/0165-1684(95)00047-H

5. Bellman, R.: On the application of dynamic programing to the determination of optimal play in chess and checkers. Proc. Natl. Acad. Sci. U.S.A. **53**(2), 244 (1965)

6. Brayton, R.K., Sangiovanni-Vincentelli, A.L., McMullen, C.T., Hachtel, G.D.: Logic Minimization Algorithms for VLSI Synthesis. Kluwer Academic Publishers, Norwell (1984)

7. Bryant, R.E.: Graph-based algorithms for Boolean function manipulation. IEEE Trans. Comput. **35**(8), 677–691 (1986)

8. Campbell, M., Hoane, A.J., Jr., Hsu, F.H.: Deep blue. Artif. Intell. **134**(1–2), 57–83 (2002). https://doi.org/10.1016/S0004-3702(01)00129-1

9. Coudert, O.: Two-level logic minimization: an overview. Integr. VLSI J. **17**(2), 97–140 (1994). https://doi.org/10.1016/0167-9260(94)00007-7

10. Culberson, J.C., Schaeffer, J.: Pattern databases. Comput. Intell. **14**(3), 318–334 (1998). https://doi.org/10.1111/0824-7935.00065

11. Damodare, R.P., Augustine, J., Jacob, J.: Lossless and lossy image compression using Boolean function minimization. Sadhana **21**(1), 55–64 (1996). https://doi.org/10.1007/BF02781787

12. Edelkamp, S., Kissmann, P.: On the complexity of BDDs for state space search: a case study in connect four. In: Proceedings of the Twenty-Fifth AAAI Conference on Artificial Intelligence, AAAI 2011, San Francisco, California, USA, 7–11 August 2011. AAAI Press (2011). http://www.aaai.org/ocs/index.php/AAAI/AAAI11/paper/view/3690

13. Falsinelli, B., Dart, J., de Man, R.: Fathom (2015). https://github.com/jdart1/Fathom

14. Fišer, P., Hlavička, J.: Implicant expansion methods used in the BOOM minimizer (2001)

15. Fišer, P., Rucký, P., Váňová, I.: Fast Boolean minimizer for completely specified functions, pp. 122–127 (2008)

16. Fišer, P., Toman, D.: A fast SOP minimizer for logic functions described by many product terms, pp. 757–764 (2009)

17. Gasser, R.: Solving nine men's morris. Comput. Intell. **12**(1), 24–41 (1996). https://doi.org/10.1111/j.1467-8640.1996.tb00251.x

18. Heinz, E.: Endgame databases and efficient index schemes for chess. J. Int. Comput. Games Assoc. **22**(1), 22–32 (1999)

19. van den Herik, H., Herschberg, I.: The construction of an omniscient endgame data base. J. Int. Comput. Chess Assoc. **8**(2), 66–87 (1985)

20. Hong, S.J., Cain, R.G., Ostapko, D.L.: MINI: a heuristic approach for logic minimization. IBM J. Res. Dev. **18**(5), 443–458 (1974). https://doi.org/10.1147/rd.185.0443

21. killrducky: TB rescoring (2018). https://blog.lczero.org/2018/09/tb-rescoring.html

22. de Man, R.: Syzygy (2020). https://github.com/syzygy1/tb

23. Nalimov, E.V., Haworth, G.M., Heinz, E.A.: Space-efficient indexing of chess endgame tables. J. Int. Comput. Games Assoc. **23**(3), 148–162 (2000)

24. v. Neumann, J.: Zur theorie der gesellschaftsspiele ("On the theory of parlor games"). Math. Ann. ("Math. Ann.") **100**, 295–320 (1928). https://doi.org/10.1007/BF01448847

25. Newell, A., Shaw, J.C., Simon, H.A.: Chess-playing programs and the problem of complexity. IBM J. Res. Dev. **2**(4), 320–335 (1958). https://doi.org/10.1147/rd.24.0320

26. Patashnik, O.: Qubic: $4 \times 4 \times 4$ tic-tac-toe. Math. Mag. **53**(4), 202–216 (1980). https://doi.org/10.1080/0025570X.1980.11976855

27. Quinlan, J.R.: Learning efficient classification procedures and their application to chess end games. In: Michalski, R.S., Carbonell, J.G., Mitchell, T.M. (eds.) Machine Learning: An Artificial Intelligence Approach. Symbolic Computation, vol. 1064, pp. 463–482. Springer, Heidelberg (1983). https://doi.org/10.1007/978-3-662-12405-5_15

28. Saffidine, A., Jouandeau, N., Buron, C., Cazenave, T.: Material symmetry to partition endgame tables. In: van den Herik, H.J., Iida, H., Plaat, A. (eds.) CG 2013. LNCS, vol. 8427, pp. 187–198. Springer, Cham (2014). https://doi.org/10.1007/978-3-319-09165-5_16

29. Sarkar, D.: Boolean function-based approach for encoding of binary images. Pattern Recogn. Lett. **17**(8), 839–848 (1996). https://doi.org/10.1016/0167-8655(96)00045-1

30. Schaeffer, J., et al.: Checkers is solved. Science **317**(5844), 1518–1522 (2007). https://doi.org/10.1126/science.1144079. https://science.sciencemag.org/content/317/5844/1518

31. Shannon, C.E.: A chess-playing machine. Sci. Am. **182**(2), 48–51 (1950). https://doi.org/10.1038/scientificamerican0250-48

32. Silver, D., et al.: A general reinforcement learning algorithm that masters chess, shogi, and Go through self-play. Science **362**(6419), 1140–1144 (2018). https://doi.org/10.1126/science.aar6404

33. Stiller, L.: Parallel analysis of certain endgames. J. Int. Comput. Chess Assoc. **12**(2), 55–64 (1989)

34. Stiller, L.: Group graphs and computational symmetry on massively parallel architecture. J. Supercomput. **5**(2–3), 99–117 (1991). https://doi.org/10.1007/BF00127839

35. Stiller, L.: Multilinear algebra and chess endgames. In: Games of on Chance, vol. 29, pp. 151–192 (1996)

36. Ströhlein, T.: Untersuchungen über Kombinatorische Spiele ("Investigations on combinatorial games"). Ph.D. thesis, Fakultät für Allgemeine Wissenschaften der Technische Hochschule München ("Faculty of General Sciences of Munich Technical University") (1970)

37. Thompson, K.: Retrograde analysis of certain endgames. J. Int. Comput. Chess Assoc. **9**(3), 131–139 (1986)

38. Turing, A.M.: Digital computers applied to games. Faster than Thought (1953)

39. Yang, J., Savari, S.A., Mencercv, O.: Lossless compression using two-level and multilevel Boolean minimization. In: 2006 IEEE Workshop on Signal Processing Systems Design and Implementation, pp. 148–152 (2006)

40. Zakharov, V., Maknhychev, V.: Creating tables of chess 7-piece endgames on the Lomonosov supercomputer. Superkomp'yutery ("Supercomputers") **15**, 34–37 (2013)

41. Zermelo, E.: Über eine anwendung der mengenlehre auf die theorie des schachspiels ("On an application of set theory to the theory of the game of chess"). In: Proceedings of the Fifth International Congress of Mathematicians, vol. 2, pp. 501–504 (1913)

Player Modelling

Procedural Maze Generation Considering Difficulty from Human Players' Perspectives

Keita Fujihira[(✉)], Chu-Hsuan Hsueh, and Kokolo Ikeda

School of Information Science, Japan Advanced Institute of Science and Technology, Nomi, Ishikawa, Japan
{keita.fujihira,hsuehch,kokolo}@jaist.ac.jp

Abstract. In video game development, creating maps, enemies, and many other elements of game levels is one of the important processes. In order to improve players' game experiences, game designers need to understand players' behavioral tendencies and create levels accordingly. Among various components of levels, this paper targets mazes and presents an automatic maze generation method considering difficulty based on human players' tendencies. We first investigate the tendencies using supervised learning and then create a test player considering human-likeness by exploiting the tendencies. The test player simulates human players' behaviors when playing mazes and judges difficulty according to the simulation results. Maze evaluation results from subject experiments show that our method succeeds in generating mazes where the difficulty estimated by the test player matches human players'.

Keywords: Procedural content generation · Maze · Difficulty · RPG

1 Introduction

Artificial Intelligence (AI) has succeeded in various fields; for games, AI has been applied not only to make computer players but also to generate game content, well known as Procedural Content Generation (PCG). PCG has attracted attention from academia and industry and was mainly researched for popular game genres such as platformer games (e.g., Super Mario Bros) [1] and shooting games [2]. In contrast, research on role-playing games (RPG) is relatively few.

RPG is a classical genre that includes famous titles such as The Elder Scrolls[1] and Fallout[2] series. In addition to defeating final bosses, players may have various purposes, and these factors are often reflected in game design. One example is "finding and collecting items on the maps." Since it is not exciting that rare items are easily found, game designers may design maps that guide players away

---

[1] https://elderscrolls.bethesda.net/en.
[2] https://fallout.bethesda.net/en/.

© Springer Nature Switzerland AG 2022
C. Browne et al. (Eds.): ACG 2021, LNCS 13262, pp. 165–175, 2022.
https://doi.org/10.1007/978-3-031-11488-5_15

from rare items. Another example is "going to the next town." Although exploring game maps may be enjoyable, getting lost for a long time is frustrating. Meanwhile, too explicit guidance (e.g., maps containing a single path or flashing marks showing the correct directions) harms the play experience. Therefore, it is better that RPG maps give players freedom in exploring while implicitly controlling their behaviors. These are key challenges in game design, known as the "Narrative Paradox" [3]. Creating desirable maps usually needs many game designers' efforts and time, making the development costs high.

The main goal of this research is the automatic generation of such maps. As the first step, we target simple two-dimensional mazes that contain only passages and walls but no enemies, items, or NPCs. We investigate human players' behavioral tendencies when playing such mazes. In this paper, we aim to know the difficulty of mazes from human players' perspectives and generate mazes accordingly. We analyze human players' path selection tendencies by supervised learning and use the learned model to create a test player for maze generation. In evaluation experiments, the results that human players play the generated mazes are generally consistent with the results predicted by the test player.

The rest of the paper is organized as follows. Section 2 introduces work related to PCG and maze generation. Section 3 and Sect. 4 present an investigation of human players' tendencies using supervised learning and a maze generation approach using the test player, respectively. Section 5 shows the results of the subject experiments on maze evaluation. Finally, Sect. 6 makes conclusions.

2   Related Work

PCG is the algorithmic creation of game content with limited or indirect user input [4]. A large variety of content such as maps [5], puzzles [6], and NPCs [7] is covered as targets of PCG. Among approaches for PCG, the following four are representative. Constructive PCG generates content via rules that are usually hand-crafted. Search-based PCG optimizes generated content through repeats of generation and evaluation [8]. PCG via Machine Learning trains generation models using existing game content to generate new one [9]. PCG via Reinforcement Learning is a very recent approach that trains generation models by reinforcement learning and does not require existing game content [10,11].

This paper focuses on generating maze maps. Simple Constructive PCG algorithms include digging [12], extending [13], and toppling [14] methods. Some researchers further considered maze difficulty and used Search-based PCG [15–17]. For example, Kwiecień [17] proposed to generate challenging mazes based on the cockroach swarm optimization algorithm and defined maze complexity by the length of a solution path, the number of its direction changes, and the branch point number in the maze. However, none of them explicitly considered human players' behaviors. Even though a maze looks complicated containing many branch points, players do not always get lost. Our approach differs from these papers' in that ours considers difficulty from human players' perspectives.

3   Investigation of Human Players' Tendencies

Mazes have been widely played around the world, which can be further grouped by different attributes such as the number of dimensions and the uniqueness of the solution path. In this paper, we target two-dimensional mazes with unique solution paths. In order to adjust difficulty from humans' perspectives, it is important to grasp human players' behavior tendencies. Section 3.1 presents subject experiments on collecting human players' behaviors. Section 3.2 employs supervised learning on path selection probability to investigate the tendencies.

3.1   Subject Experiments

We generated the mazes by classical algorithms [12–14]. Figure 1(a) and 1(b) show the screenshots of mazes, and the maze settings are listed as follows:

- The maze size is 31×31 (Small), 41×41 (Medium), or 51×51 (Large).
- The maze consists of only passable cells (passage) and impassable cells (wall).
- Since the algorithms place two consecutive passages at a time, each passage must locate at a cell whose x-, y-, or both coordinates are even numbers, assuming that the top-left corner is (1, 1).
- A player's goal is to get from a start point to an end point.
- In cases that players do not visit a cell more than once, the solution path that connects the start point to the end point is unique.
- The start point and the end point are located at the upper left and lower right in the maze, respectively.
- The maze does not contain loops, cycles, and isolated cells.
- A player can move only one cell vertically or horizontally per action.
- The time limit is set for playing a maze, depending on the maze size: 80 s for Small, 100 s for Medium, 150 s for Large. If the player does not reach the end point within the time limit, the game is over.
- A player has either one of the following two ranges of view.
  - Wide view: A player can view the entire maze, as shown in Fig. 1(a).
  - Narrow view: A player can only view inside a circle centered on him-self/herself, as shown in Fig. 1(b). The diameter of the circle is half of the side length of the maze (e.g., $31/2 = 15.5$ cells for Small)[3].

A total of twenty players (males and females in their twenties to forties) participated in the experiments. The participants consisted of both players who were interested in playing games and players who were not. The participants played 21 mazes with different maze sizes and view ranges.

---

[3] We decide the circle size considering both humans' visual perception and the design of RPG maps. We leave it as future investigations into whether the sizes fit human players' visual perception and how different sizes influence the results.

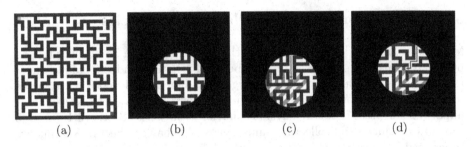

Fig. 1. Small size mazes with (a) the wide view and (b) the narrow view, and intersections that (c) is not and (d) is a branch point, where the yellow, pink, and red cells are the start point, the end point, and the player, respectively. (Color figure online)

3.2 Prediction by Supervised Learning

Human players' tendencies are investigated by employing supervised learning on probabilities of selecting proceeding directions at *branch points*. Note that not all intersections are branch points, which we will define soon later. For a given cell in a maze, we define a proceeding direction as *uncertain* if any part of the succeeding paths is invisible within the narrow view range. In Fig. 1(c), if the player (red cell) goes right, it leads to succeeding paths in green oblique lines. Since some paths go to areas outside of the narrow view, the proceeding direction of right is uncertain. In contrast, going left leads to succeeding paths fully visible in the narrow view (blue vertical lines) and is not uncertain. We then define a *branch point* as an intersection connected to two or more uncertain proceeding directions. In Fig. 1(d), the player (red cell) is at a branch point connected to two uncertain proceeding directions. Note that branch points in both wide and narrow view mazes are defined in the same way.

Learning Settings. We collected branch points and the players' selection proportions of proceeding directions from the mazes in the subject experiments. For example, assume that players could go right or go down at some branch point. If 16 out of 20 players went right, the selection proportion of going right at this branch point was $16/20 = 0.8$. Note that we only counted data from the first time that players reached the branch points. Also, we only considered uncertain proceeding directions. For the supervised learning model, the input was maze features related to branch points, including a proceeding direction, and the output was the predicted selection proportion. We extracted 23 input features, as explained in Appendix A. We collected 147 and 197 training data from the subject experiments for wide and narrow view mazes, respectively. Since the amount of data was small, we doubled the data by creating copies that swapped the x-axis and y-axis values of the original maze data. The prediction models were built based on the LightGBM[4] with leave-one-out cross-validation.

---

[4] https://lightgbm.readthedocs.io/en/latest/.

Learning Results. Figure 2 shows the prediction results. Note that selection proportions of the proceeding directions at a branch point were predicted separately and that the sum of the proportions might not be 1.0. Thus, we further normalized the selection proportions at each branch point to make them a probability distribution (i.e., summing to 1.0). Root-mean-square errors between the predicted probabilities and actual proportions were 0.26 (wide view) and 0.21 (narrow view). Although the prediction accuracy still had room to improve, the Pearson correlation coefficients were 0.66 (wide view) and 0.74 (narrow view). The former had a moderate positive correlation, while the latter had a highly positive correlation. We concluded that the prediction models were reliable to some extent, especially the narrow view one.

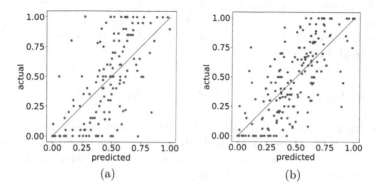

Fig. 2. Prediction results (x-axis: predicted probability, y-axis: actual proportion) of (a) wide view mazes and (b) narrow view mazes.

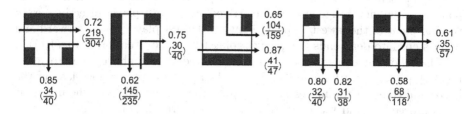

Fig. 3. Human players' selection proportions of the proceeding directions.

We further analyzed the players' tendencies in both wide view and narrow view mazes and had interesting findings reflected in the prediction models. Due to the page limit, we only present the results of narrow view mazes. We observed that (1) the players tended to go down or right instead of going up or left and that (2) when both right and down directions were available, the players tended to go straight. Such tendencies are shown in Fig. 3, the statistics on the players' selection proportions of proceeding directions according to the shapes of branch points. To keep the figures simple and easy-to-read, data with selection proportions less than 0.5 are excluded. Also, data with sample numbers less than 20 are excluded since they are not reliable enough.

We considered that tendency (1) is because the end points of all mazes in the experiments were fixed at the bottom-right corner. When further looking into the prediction model, the "cos_goal" feature had the biggest contribution to prediction (feature importance of 0.35). With a higher "cos_goal" value, the proceeding direction was likely promising for the players. For example, if a player is at some branch point in the upper-right area, going down leads to the highest "cos_goal" value and is reasonable for the player to select. Thus, we concluded that tendency (1) was adequately reflected in the prediction model. As for tendency (2), we considered that going straight is easier to operate than changing directions for human players. This tendency was also reflected in the prediction model, where the "is_straight" feature contributed the second most (feature importance of 0.15).

4   Maze Generation and Selection Approach

For generated mazes, we aim to evaluate the difficulty from human players' perspectives. With a prediction model (e.g., the narrow view one in Sect. 3), Sect. 4.1 creates a test player simulating human players. Section 4.2 then defines difficulty and generates mazes accordingly.

4.1   Test Player Considering Human-Likeness

The test player's action selection is broadly divided into two cases according to whether the passage is a branch point or not (defined in Sect. 3.2). For a passage that is not a branch point, the test player proceeds as follows. When the end point is in the view range and can be arrived at, the test player directly goes to the end point in the fewest steps. Otherwise, the test player moves forward and never steps into dead ends (e.g., never going left in the situation of Fig. 1(c)).

The following two exceptions are introduced to the above rules to make the test player more human-like. First, when the test player goes left on the horizontal edges of the mazes or goes up on the vertical edges of the mazes, it stops proceeding and returns to the last passed branch point. Since all mazes have end points fixed at the bottom-right corner and do not contain loops, going left or up on the edges never leads to solution paths. The investigation in Sect. 3 also supports that human players have such a tendency. Second, for narrow view mazes, the test player returns to the last passed branch point upon knowing that the succeeding paths lead to dead ends.

For a branch point, the test player only selects uncertain proceeding directions, and the selection is different from whether it is the first time visiting or not. When visiting a branch point for the first time, the test player selects directions according to the probability distribution from the prediction models.

When the test player returns to a branch point from some dead ends, we predict the selection proportions again for directions that the paths have not been traversed. When the test player is not on the solution path[5], the predicted proportions less than 0.5 are changed to 0.0, which is to prevent the test player's behaviors from being too different from human players'. If all the not-traversed directions have proportions of 0.0, the test player returns to the last passed branch point. The reason for not applying this rule to solution paths is to make sure that the test player can reach the end points. If at least one direction can be selected, the proportions are normalized to sum to 1.0 no matter whether the branch point is on the solution path or not. Taking a T-shaped branch point on the solution path as an example, assume that the predicted proportions for going right and down are 0.6 and 0.3, respectively. The test player goes right with a probability of $0.6/(0.6 + 0.3) \approx 67\%$ and goes down with 33%.

4.2 Maze Difficulty Evaluation Using Test Player

We use the test player to simulate human players' behaviors when playing mazes and evaluate difficulty according to estimated step numbers. In more detail, for each maze, we know in advance the shortest number of steps to the end point, denoted by n^*. After the test player clears the maze, we count the total number of steps, denoted by n_{total}. We then define *the number of extra steps* n_{extra} as $n_{total} - n^*$. With higher n_{extra}, we consider that the maze is more difficult. Since the test player involves randomness, we let it play each maze 10 times and judge the difficulty according to the results of all trials. We generate mazes automatically using the digging method [12] and define difficulty by two indicators, average n_{extra} and the standard deviation of n_{extra}, as: easy/moderate/difficult whose average n_{extra} is low/moderate/high and lowSD/highSD whose standard deviation of n_{extra} is low/high. For example, difficult$_{lowSD}$ contains mazes with a high average n_{extra} and a low standard deviation of n_{extra}.

5 Subject Experiments on Maze Evaluation

We conducted subject experiments to see whether the maze difficulty for human players is well predicted by the test player. A total of 10 players (males in their twenties) participated in the experiments. The participants were different from those in Sect. 3 for the sake of fair evaluation. The maze size was 51×51, and the players had the narrow view. We generated 30,000 mazes and let the test player with the narrow view prediction model play each maze 10 times. A maze with an average n_{extra} in $[0, 50)$ was classified as easy, $[150, 250)$ as moderate, and $[350,)$ as difficult. In each class, mazes with the top-7 and the bottom-7 standard deviations of n_{extra} were selected as highSD and lowSD, respectively. We excluded easy$_{highSD}$ since the standard deviations did not differ too much from easy$_{lowSD}$. Therefore, we had five difficulty groups, as shown in Table 1.

---

[5] We create the test player assuming that the shortest solutions of mazes are known.

Figure 4(a) and 4(b) show examples of easy$_{\text{lowSD}}$ and difficult$_{\text{lowSD}}$ mazes with the test player's trajectories of one trial, where cells in gray are on the solution path and those in red oblique lines are not. Figure 4(c) shows the trajectories by one of the human players, which looks particularly similar to the test player's.

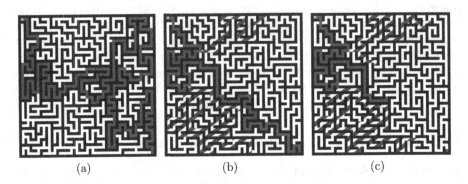

| | (a) | | (b) | | (c) |

Fig. 4. Example mazes of (a) easy$_{\text{lowSD}}$ and (b)(c) difficult$_{\text{lowSD}}$, where (a) and (b) show the test player's trajectories in gray (on the solution path) and in red oblique lines (not on the solution path), and (c) shows one human player's trajectories. (Color figure online)

Table 1. The means and standard deviations (SD) of n_{extra} by human players.

easy$_{\text{lowSD}}$			moderate$_{\text{lowSD}}$			moderate$_{\text{highSD}}$			difficult$_{\text{lowSD}}$			difficult$_{\text{highSD}}$		
order	mean	SD	order	mean	SD	order	mean	SD	order	mean	SD	order	mean	SD
5	200	194	1	71	43	3	53	113	4	299	173	2	257	180
9	72	131	10	241	197	6	243	206	7	364	123	8	231	162
14	69	44	12	285	114	11	195	209	13	371	121	17	447	162
18	77	116	19	220	137	15	135	214	16	406	150	20	368	200
21	175	151	23	228	140	24	232	201	22	262	133	25	265	234
26	48	47	29	178	152	27	228	180	28	517	159	30	104	173
33	27	25	32	235	42	31	314	291	34	374	77	35	140	184
Avg.	95	101	Avg.	208	118	Avg.	200	202	Avg.	307	134	Avg.	259	185

Table 1 shows the order that the participants played the 35 mazes as well as the means and the standard deviations of the participants' n_{extra} for each maze. Generally, the difficulty judged by the test player matched human players' (i.e., difficult > moderate > easy). When focusing on mazes with lowSD, the Student's t-test ($p < 0.05$) showed that the results were statistically significant. Namely, difficult$_{\text{lowSD}}$ had the average n_{extra} higher than moderate$_{\text{lowSD}}$ ($p = 0.0016$), and moderate$_{\text{lowSD}}$ higher than easy$_{\text{lowSD}}$ ($p = 0.0083$). We also compared the standard deviations within the same difficulty. We considered that mazes with lower standard deviations are preferred since it means that even different human players may still experience similar difficulty. For mazes with lowSD and highSD, the results of Student's t-test ($p < 0.05$) showed statistical significance as follows: moderate$_{highSD}$ had the average standard deviation higher than moderate$_{lowSD}$ ($p = 0.0142$), and difficult$_{highSD}$ higher than difficult$_{lowSD}$ ($p = 0.0057$). The results demonstrated that our method succeeded in generating mazes where the difficulty matched human players' in terms of n_{extra}.

Interestingly, some easy$_{lowSD}$ mazes were actually difficult for human players and vice versa, as shown in Table 1. We suspected that the accuracy of the prediction model was insufficient in some situations and the test player somewhat lacked the consideration for humans' assumptions and misunderstandings. For example, assume a maze has a very long solution path. When human players cannot reach the end point for a long time, they may return to previous branch points even when they are on the solution path. In contrast, the test player did not have such hesitation. We expected that the difficulty evaluation would fit human players' behaviors better if we introduced more human-like characteristics like this into the test player.

6 Conclusions and Future Work

In this paper, we proposed a procedural maze generation method considering difficulty from human perspectives, which generated mazes by the following two steps. The first investigated human players' tendencies. We employed supervised learning and built models for predicting human players' path selection probabilities. The prediction results had moderate or highly positive correlations to human players' selections. The second step created a test player based on the model. The test player's results of playing mazes were used as a measure of difficulty. Then, we conducted subject experiments to evaluate whether the maze difficulty is suitable for human players. The experiments showed that difficulty estimated by the test player matched human players' playing results in general.

The followings discuss several promising research directions. About the test player, on the one hand, we expect to improve the difficulty evaluation by making the test player consider more human-like characteristics. On the other hand, to make our approach applicable to a wide variety of mazes, it is worth investigating more general ways to create human-like test players. Also, the efficiency of maze generation has room to improve. The current approach may generate many mazes before obtaining one with a specific difficulty. We consider that combining other algorithms such as simulated annealing and local search can help improve the efficiency. In addition, we plan to compare our approach to existing ones in terms of difficulty evaluation, investigate the influences from the size of narrow view or different locations of the start point and the end point, and apply the approach to generate mazes containing more elements such as enemies and items for application to RPGs.

Acknowledgements. This work was supported by JSPS KAKENHI Grant Numbers JP18H03347 and JP20K12121.

Appendix A   Input Features of Prediction Models

The following are 23 features, mainly related to each branch point b. Directions are represented by integers, 0 for right, 1 for down, 2 for left, and 3 for up.

(1) **maze_size:** The maze size (0 for Small, 1 for Medium, or 2 for Large). (2)(3) **x, y:** The x- and y-coordinates of b. (4) **ent:** The direction from which players enter b. (5) **proc:** The proceeding direction at b. (6)–(9) **proc_↑, proc_↓, proc_→, proc_←:** Whether each direction is an uncertain proceeding direction. (10)–(13) **dist_edge_↑, dist_edge_↓, dist_edge_→, dist_edge_←:** Distance from b to the edge in each direction. (14) **is_straight:** Whether ent and proc are in a straight line. (15) **straight_depth:** The number of passages that follow the straight line. (16) **promisingness:** The number of passages on the succeeding paths of proc regardless of the view range. (17) **promisingness_sum:** The sum of the promisingness values of all directions, excluding ent, at b. (18) **general_direction:** The general direction of the paths extending from proc. (19) **num_branch:** The number of other branch points in the narrow view range, which was used for both wide and narrow view mazes. (20) **dist_start:** The minimum number of steps from the start point to b. (21) **avg_step:** The number of steps taken to reach b. We used the average number of steps of the experiment participants when building the prediction models. (22)(23) **cos_start, cos_goal:** The $\cos\theta$ of the angle between the start point or end point and proc.

References

1. Volz, V., et al.: Evolving mario levels in the latent space of a deep convolutional generative adversarial network. In: 2018 Genetic and Evolutionary Computation Conference, pp. 221–228 (2018)
2. Cardamone, L., Yannakakis, G.N., Togelius, J., Lanzi, P.L.: Evolving interesting maps for a first person shooter. In: Di Chio, C., et al. (eds.) EvoApplications 2011. LNCS, vol. 6624, pp. 63–72. Springer, Heidelberg (2011). https://doi.org/10.1007/978-3-642-20525-5_7
3. Louchart, S., Aylett, R.: Solving the narrative paradox in VEs – lessons from RPGs. In: Rist, T., Aylett, R.S., Ballin, D., Rickel, J. (eds.) IVA 2003. LNCS (LNAI), vol. 2792, pp. 244–248. Springer, Heidelberg (2003). https://doi.org/10.1007/978-3-540-39396-2_41
4. Togelius, J., Kastbjerg, E., Schedl, D., Yannakakis, G.N.: What is procedural content generation?: Mario on the borderline. In: 2nd International Workshop on Procedural Content Generation in Games, pp. 1–6 (2011)
5. Togelius, J., et al.: Multiobjective exploration of the starcraft map space. In: 2010 IEEE Conference on Computational Intelligence and Games, pp. 265–272 (2010)
6. Oikawa, T., Hsueh, C.H., Ikeda, K.: Improving human players' T-spin skills in Tetris with procedural problem generation. In: Cazenave, T., van den Herik, J., Saffidine, A., Wu, I.C. (eds.) ACG 2019. LNCS, vol. 12516, pp. 41–52. Springer, Cham (2020). https://doi.org/10.1007/978-3-030-65883-0_4
7. Soares, E.S., Bulitko, V.: Deep variational autoencoders for NPC behaviour classification. In: 2019 IEEE Conference on Games, pp. 1–4 (2019)

8. Togelius, J., Yannakakis, G.N., Stanley, K.O., Browne, C.: Search-based procedural content generation: a taxonomy and survey. IEEE Trans. Comput. Intell. AI Games **3**(3), 172–186 (2011)

9. Summerville, A., et al.: Procedural content generation via machine learning (PCGML). IEEE Trans. Games **10**(3), 257–270 (2018)

10. Nam, S., Ikeda, K.: Generation of diverse stages in turn-based role-playing game using reinforcement learning. In: 2019 IEEE Conference on Games, pp. 1–8 (2019)

11. Khalifa, A., Bontrager, P., Earle, S., Togelius, J.: PCGRL: procedural content generation via reinforcement learning. In: 16th AAAI Conference on Artificial Intelligence and Interactive Digital Entertainment, pp. 95–101 (2020)

12. Algoful. https://algoful.com/Archive/Algorithm/MazeDig. Accessed Sept 2021

13. Algoful. https://algoful.com/Archive/Algorithm/MazeExtend. Accessed Sept 2021

14. Algoful. https://algoful.com/Archive/Algorithm/MazeBar. Accessed Sept 2021

15. Susanto, E.K., Fachruddin, R., Diputra, M.I., Herumuti, D., Yunanto, A.A.: Maze generation based on difficulty using genetic algorithm with gene pool. In: 2020 International Seminar on Application for Technology of Information and Communication, pp. 554–559 (2020)

16. Adams, C., Louis, S.: Procedural maze level generation with evolutionary cellular automata. In: 2017 IEEE Symposium Series on Computational Intelligence, pp. 1–8 (2017)

17. Kwiecień, J.: A swarm-based approach to generate challenging mazes. Entropy **20**(10), 762 (2018)

Opponent Model Selection Using Deep Learning

Hung-Jui Chang[1], Cheng Yueh[2], Gang-Yu Fan[3], Ting-Yu Lin[3], and Tsan-sheng Hsu[3(✉)]

[1] Department of Applied Mathematics, Chung Yuan Christian University, Taoyuan, Taiwan
[2] Department of Computer Science and Information Engineering, National Taiwan University, Taipei, Taiwan
[3] Institute of Information Science, Academia Sinica, Taipei, Taiwan
tshsu@iis.sinica.edu.tw

Abstract. It is observed that the strength of many game programs varies a lot in a tournament when facing opponents of different styles. In order to adapt one's playing strategy when playing against different programs to obtain the best possible outcome, it is important to estimate the strength of the opponent. We use a neural network to predict the strength or style of an opponent via the first 20 plies of a game in Chinese Dark Chess. According to the prediction result, a contempt factor is used when playing against the target opponent after the first 20 plies. Both the experiment result and the tournament results show that this approach can improve the rank of our program in a competitive tournament.

1 Introduction

In computer games, predicting the strength of the opponent player is one of the essential techniques to improve the level of the computer programs [7,8]. Historical data can be used to estimate a specific opponent's strength [8]. Moreover, one can use the estimated results to implement a particular strategy when playing against the target opponent [6,7].

One of the strategies used in computer Chess in finding a good opponent model is called the *contempt factor* [7], which is a value indicating the strength difference between two programs. When this value is non-negative, we assume the opponent has an equal or greater strength than ours. In this case, we accept a draw even if we currently have a small advantage. On the other hand, if the contempt factor is negative, the opponent is assumed to be weaker than ours. In this case, we do not accept a draw even if we currently have a small disadvantage. That is, we try to force a draw when the opponent is stronger and avoid a draw result when the opponent is weaker.

There are many participants in a computer game tournament [3]. Some participants have participated in the tournament several times, but there are always

This work was supported in part by the Ministry of Science and Technology of Taiwan under contracts MOST108-2221-E-001-011-MY3 and MOST110-2222-E-033-005-.

© Springer Nature Switzerland AG 2022
C. Browne et al. (Eds.): ACG 2021, LNCS 13262, pp. 176–186, 2022.
https://doi.org/10.1007/978-3-031-11488-5_16

some newcomers. We can use the previous results to estimate the strengths of those who have participated in the tournament before. For newcomers, we need to design a method to predict their strengths. Knowing that your opponent is weak, we can afford to sacrifice the top-ranked piece and eventually win the game in Chinese Dark Chess, whose goal is to capture all of the opponent's pieces. This is particularly true for Chinese Dark Chess, which easily draws by chasing the opponent's most important piece, namely King, using the least valued piece, namely Pawn. Figure 1 shows an example of a forced draw. When the black pawn moves from B5 to C5, the red king's only escape is to move from C4 to B4. The black pawn can continue chasing the red king by moving from C5 to either C4 or B5. Furthermore, Chinese Dark Chess is a non-deterministic game. By flipping a dark (unrevealed) piece, a game may turn from advantage to disadvantage. For example, in Fig. 2a, the red side may consider flipping the unrevealed piece in C5. If the revealed piece is a black cannon, as in Fig. 2b, the red king can capture one black cannon and two black guards, which is a huge advantage. On the other hand, if the revealed piece is a black pawn, as in Fig. 2c, the red king will be captured directly, which is a huge disadvantage. In this paper, we provide a way to estimate the opponent's program's strength by feeding the first 20 plies of a game into a deep-learning neural network model [4,5].

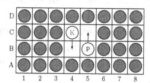

Fig. 1. An example of using black pawn to chase red king. (Color figure online)

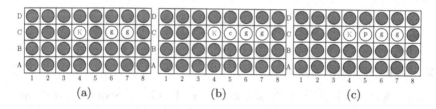

(a)   (b)   (c)

Fig. 2. An example of revealing result causes advantage and disadvantage result. (Color figure online)

In this paper, we use a deep learning network to predict the opponent's strength based on the first 20 plies of the game. The training data are collected from the previous TAAI, TCGA, and ICGA tournaments, and the self-played data from three programs with different strengths. The experiment results show that the opponent prediction model can improve the expected score of our program in a competitive tournament.

The remains of this paper is organized as follows: In Sect. 2, we briefly introduce the game of Chinese Dark Chess and the data sets of the deep-learning models. In Sect. 3, we describe the predicting methods. In Sect. 4, we show the experiment results. Finally, In Sect. 5, we conclude this paper.

2   Background

In this section, we first introduce the terminologies used in this paper. Then we describe the game *Chinese Dark Chess* which our program plays. Finally, we describe the background of our problem and the main idea of how to use the contempt factor in solving our problem.

2.1   Terminologies

A *board position* (b) denotes the state of a game. A *legal ply* (p) is an action that can transfer the current board position into the next board position, that is, $T(b, p) = b'$, where b' is the board position after ply p is applied to board position b. The *ply sequence* of a game is represented as $p_1, p_2, \ldots, p_n$, where p_i is the i-th ply of the game. The *board position sequence* of a game is represented as $b_0, b_1, \ldots, b_n$, where b_i denotes the board position after ply p_i is applied on board position b_{i-1}. An example of the relationship between the board positions and the moves is shown in Fig. 3. The *opponent ply sequence* (OPS) is the sequence of all the opponent's plies. If the opponent makes the first ply in the game, then the OPS is $\{p_1 p_3 p_5 \ldots\}$, and if we make the first ply in the game, then the OPS is $\{p_2 p_4 p_6 \ldots\}$, respectively.

Fig. 3. An illustration of board positions and ply sequences.

2.2   Chinese Dark Chess

Chinese Dark Chess (CDC) [2], also known as Banqi, is one of the most popular variations of *Chinese Chess* (CC) in southeast Asia. This game is played by the same pieces set on the half part of the original game board of CC. There are two colors in the CDC, that is, the red side and the black side. Each side has 16 pieces, one king (K/k), two guards (G/g), two ministers (M/m), two rooks (R/r), two knights (N/n), two cannons (C/c), and five pawns (P/p). In the following, we use the uppercase character to represent the pieces of the red side and the lowercase character to represent the pieces of the black side. Only a piece with a higher rank can capture an opponent piece of lower rank. The order between two pieces, which is used to decide which one can capture the other, is described by three rules: 1) $K > G > M > R > N > C$, 2) $P > K$ and 3) $G, M, R, N > P$.

Each piece in the CDC has two faces: the front-face and the back-face. The front-face of each piece is crafted with the name of that piece. The back-face of all the pieces is identical.

At the beginning of the game, all the pieces are shuffled and placed with the front-face facing-down. An illustration of the start position of the CDC is shown in the left of Fig. 4. In the beginning of the game, the first player chooses one facing-down piece and reverses it. In the remains of the game, the first player owns those pieces with the same color as the first revealed one, and the second player owns pieces of the other color.

In each turn, one player can either move a piece of his/her side to the adjacent square, reveal one facing-down piece, or apply the special jump-move for cannon. A piece can only be moved to the adjacent square, which is empty or is occupied by an opponent's piece with a lower rank. In the later case, the opponent's piece is captured by our piece and removed from the game. In the right of Fig. 4, the legal moves of red minister at C3 are C3–C2 and C3–B3, since C2 is an empty square and the black rook at B3 is lower-ranked than the red minister. Note that C3–C4 is an illegal move since C4 is occupied by the red knight, which is the same side as the red minister. Also, C3–D3 is an illegal move because the black king is higher-ranked than the red minister. A *jump move* is a special capture move for the cannons. When there is only one piece in between the cannon and the target piece, the cannon can use the jump move to capture that target piece. For example, in the right of Fig. 4, the red cannon in B5 can use a jump move to capture either the black pawn in B5, the black rook in B3, or the black guard in B8. Note that for each cannon, there exists at most one jump move along the short edge, that is B5 to D5 in our example. In our example, there can be two jump moves along the long edge, B5 to B3 and B5 to B8. We denote these two moves as the jump moves along the long edge to the left and the jump moves along the long edge to the right. The player who cannot make any legal move loses the game. As a remark, if all pieces of one side are captured, he naturally cannot make any legal move.

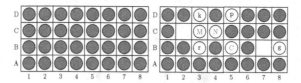

Fig. 4. The initial position of Chinese Dark Chess and an ongoing position that the legal moves of the red minister are C3–C2 and C3–B2. And the legal jump moves of the red cannon at B5 are B5–D5, B5–B3, and B5–B8. (Color figure online)

2.3   Game Tournament and Expected Score

In the competition of computer games, e.g., the World Champion of Computer Chess or Computer Olympics [3], the Swiss-system tournament is the most commonly applied playing rule. In this rule, one win gains 2 points, and one draw gains

1 point. In a tournament, we can rank other participants according to their past earned scores. We defined three different ranks: S, E, and W, that is, those who are *stronger* than us (S), those who are relatively *equal* to us (E), and those who are *weaker* than us (W). When we play against an opponent with rank x, we have a corresponding winning probability $P_{W,x}$, a losing probability $P_{L,x}$, and the probability of having a draw result with $P_{D,x}$, respectively. Let the numbers of participations with rank S, E, and W, be N_S, N_E, and N_W, respectively. The *expected score* (ES) is defined as

$$ES = \sum_{i \in \{E,S,W\}} N_i \times (2P_{W,i} + P_{D,i}). \tag{1}$$

In order to increase ES, we have two different approaches: 1) increasing $P_{W,W}$ by decreasing $P_{D,W}$; and 2) increasing $P_{D,S}$ by decreasing $P_{L,S}$. In the first approach, we try to avoid a draw when we still have a chance to win, usually this is what we want when we play against weaker opponents. In the second approach, we try to avoid a lose by using a force drawing strategy.

2.4   Contempt Factor and Threefold Repetition

In Chess and many other games, a threefold repetition is often treated as a draw. Therefore many programs return a value of 0 when a threefold repetition occurs. In Fig. 5, the left tree shows one move with a value of 0 and another move with a small positive value $+\epsilon$. The right tree shows another example: one move with a value of 0 and another move with a small negative value $-\epsilon$. The search algorithm will compare the returned value of 0 with other moves' return value. Traditionally, if there is one move whose value is greater than 0, we will not choose the move with the value of 0 no matter how small it is. However, when we play against a strong program, the position with a small positive value may not result in a win. Moreover, the risk of being beaten by a stronger opponent is high when we only have a small advantage. On the other hand, when we play against a weaker program, we will tolerate the small amount of disadvantage because we know we still have a chance to win that game even currently we only have a small disadvantage.

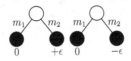

Fig. 5. An example of plies with values of 0.

The strategy described above is the main idea of using the contempt factor [7]. However, we are not sure about the strength of opponents before a tournament. For those who have participated in the tournament before, we can use its past results to estimate. For those newcomers, we need a reliable method to estimate their strength during the competition.

2.5   Chinese Dark Chess Programs

In this subsection, we describe the three CDC programs involved in our experiment. The first program is "Yahari," who participated in the computer game tournament of TAAI, TCGA, and Computer Olympic for several years, and achieved first place in TAAI 2013 and TCGA 2014, second place in 18th and 19th Computer Olympic, and the third place in TCGA 2019, TCGA 2020, and TAAI 2019. The second program is "PupilDarkChess" which achieves first place in TCGA 2020, TCGA 2019, and TAAI 2019. The third program is "YanYu," which has participated from TCGA 2018, does not win any top 3 places. It appears that "Yahari" is stronger than "YanYu" and is slightly weaker or equal to the strength of "PupilDarkChess." "Yahari" is the target program we want to improve. "Yahari" has an evaluation function with a range of -1 to 1 where -1 means loss, 1 means win, and 0 means draw.

3   Method

To determine an opponent's strength, we collect the *Opponent Ply Sequences* (OPSs) generated during the game as the input. And feed these input data into an *opponent ranking neural network* (ORNN) model to predict its rank. After obtaining the estimated strength, we set a corresponding contempt factor when playing against that program. To determine the proper value of the contempt factor, we calculate the average winning rate when "Yahari" plays against an opponent using different contempt factors.

3.1   Input Data of the ORNN

The input of our ORNN model contains one board position after the 3rd ply is played and the following seven opponent's plies. There are two reasons to use the board position after the 3rd ply is played as the initial board position. First, although the start board position of the CDC is random, there are strategies to deal with what pieces should be in the first few plies [1]. An example is when a high-ranked opponent's piece is revealed, we cannot flip pieces around that piece. These fixed strategies easily confuse a prediction program. Second, the predicted result is used to determine the strategy when playing with this opponent, that is, the contempt factor. And the contempt factor affects the search behavior of a program only when a threefold repetition occurs. A threefold repetition rarely occurs in the first 20 plies. Thus we use the first 20 plies in the OPS to predict the rank of the opponent's program without any risk.

3.2   Input Planes of the ORNN

There are two kinds of data in the input data of the ORNN: 1) the initial board position and 2) the following opponent's plies.

We use 30 planes to represent one initial board position, and each plane is a 4×8 block. The first 14 planes represent the locations of the revealed pieces, and the 15th and 16th planes represent the locations of the unrevealed pieces and the empty squares, respectively. The first 16 planes are the *piece location plane* (PLP), which denote the location information about the kind or the state (unrevealed/empty). We use 1 to denote one square under certain conditions. For example, the black king is in D3 in Fig. 4. Thus the corresponding PLP for the black king marks 1 on the square of D3. On the other hand, 0 is used when the condition is not fulfilled. Therefore all the other squares except D3 for the PLP of black king mark with 0. In Fig. 4, the plane representing the black king is the left matrix in Fig. 6. The unrevealed and empty squares planes are shown in the middle and right of Fig. 6, respectively.

0	0	1	0	0	0	0	0	1	1	0	1	1	1	1	1	0	0	0	0	0	0	0	0
0	0	0	0	0	0	0	0	1	0	0	0	1	1	1	1	0	1	0	0	0	0	0	0
0	0	0	0	0	0	0	0	1	1	0	1	1	1	1	1	0	0	0	0	0	0	0	0
0	0	0	0	0	0	0	0	1	1	1	1	1	1	1	1	0	0	0	0	0	0	0	0

Fig. 6. The input planes of the black king, the unrevealed squares and the empty squares of the board position shown in Fig. 4.

The 17th to 30th planes are the numbers of unrevealed pieces, respectively, for the two sides, the *unrevealed piece plane* (UPP). Each plane represents one kind of piece according to the canonical piece order of CDC, that is, "KGMRNCP-kgmrncp." For example, the 17th plane denotes the number of the unrevealed red king, and the 18th plane denotes the number of unrevealed red guards. The range of each value in the UPP is -1 to 1. The value -1 means the number of unrevealed pieces in that kind is minimal, that is 0. And the value 1 means the number of unrevealed pieces in that kind is maximal, that is, 1 for king, 5 for pawn, and 2 for all other kinds of pieces. If one piece has neither the minimal number nor the maximal number of pieces, the value is set to $2 \times \frac{\text{\# unrevealed pieces}}{\text{total number of pieces}} - 1$. The mapping between the number of unrevealed pieces and the value in the UPP is shown in Table 1.

Table 1. The corresponding value of unrevealed pieces of each kind.

	K/k	G/g	M/m	R/r	N/n	C/c	P/p
0	-1.0	-1.0	-1.0	-1.0	-1.0	-1.0	-1.0
1	1.0	0.0	0.0	0.0	0.0	0.0	-0.6
2		1.0	1.0	1.0	1.0	1.0	-0.2
3							0.2
4							0.6
5							1.0

The basic idea to represent a move in the CDC in our ORNN is to transform the original source-destination pair into a source-direction pair. We represent a move by the source position and how it moves to the destination move. In each round, there are eight possible moves: 1) up; 2) right; 3) down; 4) left; 5) flip an unrevealed piece; 6) use a cannon to capture an opponent's piece along the short edge; 7) use a cannon to capture an opponent's piece along the long edge to the left; 8) use a cannon to capture an opponent's piece along the long edge to the right. We use a $4 \times 8 \times 8$ block to encode the above information. For example, in the left board of Fig. 7, the red king can either move up, right, down, or left. If the red king moves up from C3 to C4, the corresponding input plane, the move-up plane, is represented as the middle matrix in Fig. 7, and the rest of the 7 move-planes are all zeros. If the red side plays B6–D6 to capture the black pawn, then the corresponding jump-right plane is recorded as the right matrix in Fig. 7, and the rest of the 7 move-planes are all zeros.

We use 56 planes, called the *opponent-ply plane* (OPP), to represent the opponent's following 7 plies from the initial position. In total, we use 86 planes to input one board position and the corresponding plies. The details of the input planes are summarized in Table 2.

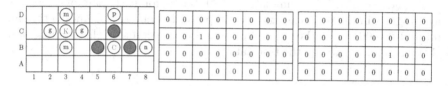

Fig. 7. The input planes of the red king's moves and the red cannon's jump-move. (Color figure online)

Table 2. Summary of the input planes.

Type	Details	# planes
PLP	Locations of revealed pieces, unrevealed pieces and empty squares	16
UPP	Number of unrevealed pieces of each kind	14
OPP	The opponent's following 7 plies	56

3.3 Output of the ORNN

As we mentioned in Sect. 2.3, an opponent's strength is estimated to be in three different levels corresponding to the S, E, and W code and in ORNN model as 0, 1, and 2, respectively.

4 Experiments

The experiment contains two parts. In the first part, we show the expected $P_{W,S}$, $P_{W,E}$ and $P_{W,L}$ when using different contempt factors. In the second part, we use the expected winning probability gathered from the first experiment to calculate the ES gain in ORNN.

4.1 Experiment Settings

In the first part of the experiment, we let our program "Yahari" play against the other two programs with three different contempt factors, −0.6, 0, and 0.2. We play 200 games for each setting.

When we have no information about the opponent's strength, we can only assume the opponent's strength is similar to our program. We use 0 as the contempt factor in this case. If our program is stronger than the opponent's, we will use a lower contempt factor. Since our program is stronger, it can gain back the advantage from slightly disadvantageous situations in most cases because a weaker makes mistakes easily. We will use −0.6 as the contempt factor in this case. That is, our program refuses to get a draw result when it still has a chance to fight back. However, when the opponent's program's strength is stronger than our program, it is hard to turn the situation from a small advantage, says board positions with evaluation value around 0.1 or 0.2, into a winning result. Therefore we accept a draw result when we only have a small advantage when playing against a stronger opponent. Hence we use 0.2 as the contempt factor.

In order to train our ORNN, we collect the game records generated by our program, "Yahari," which plays against the other two programs: "PupilDark-Chess" and "YanYu." We simulate 40,000 games for each pair of programs. Each program in the pair plays as the first player in half of the games. There are a total of 80,000 games. We use 80% of the game records as the training data and the remaining 20% as the test data.

Table 3. Yahari plays against YanYu and PupilDarkChess.

Yahari v.s. YanYu.					Yahari v.s. PupilDarkChess.				
Contempt Factor	Win	Draw	Lose	Score	Contempt Factor	Win	Draw	Lose	Score
−0.6	127	49	24	305	−0.6	42	21	137	105
0.0	123	57	20	303	0.0	41	37	122	119
0.2	120	64	8	304	0.2	48	62	90	158

4.2 The Effect of Using the Contempt Factor

In Table 3, we use three different contempt factors: −0.6, 0.0 and 0.2. For each setting, we play 200 games against "YanYu," which is a weaker program, and 200 games against "PupilDarkChess," which is a stronger program. The experiment results show that if we play against a weaker opponent, the setting contempt factor does not affect the expected score. However, the experiment results show that when using the contempt factor as −0.6, the number of draw games decreases, and the number of winning games increases. Note that when the number of draw games decreases, both the number of winning and losing games increase. That is, when playing against a weaker opponent, we prefer to have a non-draw outcome. The experiment results also show that if we play against a stronger opponent, the contempt factor helps us gain a higher expected score. When we set the contempt factor to 0.2, we force a draw when we only have a small advantage. The experiment results show a much higher expected score, 158, than the expected score of 119 as in the original setting. The experiment results also show that if one has a wrong guess and uses −0.6 as the contempt factor, it causes little harm as far as the expected score is concerned.

Table 4. Yahari v.s. Yahari.

Total	Win	Draw	Lose	Score
0 v.s. −0.6	97	23	80	217
0 v.s. 0.2	83	55	62	221
−0.6 v.s. 0.2	90	24	86	204
0 v.s. 0	77	45	78	199

In Table 4, we show the self-play result with different contempt factors. The experiment result shows that when a program with a contempt factor value of 0 plays against a program with a contempt factor value of 0.2, the one with a contempt factor 0 has a slight advantage. We believe the opponent forces a draw even when it has a small advantage. The experiment result also shows that when a program with a contempt factor value of 0 plays against a program with a contempt factor value of −0.6, the program with the contempt factor 0 also has a small advantage since the opponent tries to fight back even if they are currently behind. Therefore, when playing against the programs with a strength similar to our program, it is better to use 0 as the contempt factor.

4.3   Experiment Results

In Table 5, we show the confusion matrix of our ORNN. The experiment results show that when the opponent is weaker, equal, or stronger to our program, the prediction rate is 0.877, 0.863, and 0.855, respectively. And the overall prediction accuracy in the training data and validation data are 0.8756 and 0.8717, respectively.

Table 5. Experiment results of ORNN.

	Predict		
	W	E	S
W	0.877	0.043	0.080
E	0.046	0.863	0.091
S	0.071	0.074	0.855

5   Conclusions

This paper proposes a novel method to predict the opponent's relative ranking in the CDC tournament using the neural network. The experiment results show that the training accuracy is 0.8756 and the prediction accuracy is 0.8717. According to the experiment results, we also found the best contempt factor playing against stronger opponents is 0.2, and the contempt factor for playing against a weaker opponent is -0.6. By using the above setting, the expected score increases by 0.07 for "Yahari." By applying these methods, our program "Yahari" has improved from 3rd place in the 2019 and 2020 TCGA tournaments to 2nd place in the 2021 TCGA tournament.

References

1. Chen, B.-N., Hsu, T.: Automatic generation of opening books for dark chess. In: van den Herik, H.J., Iida, H., Plaat, A. (eds.) CG 2013. LNCS, vol. 8427, pp. 221–232. Springer, Cham (2014). https://doi.org/10.1007/978-3-319-09165-5_19
2. Chen, B.-N., Shen, B.-J., Hsu, T.: Chinese dark chess. ICGA J. **33**(2), 93–106 (2010)
3. Chen, J.-C., Lin, T.-Y., Fan, G.-Y.: Yahari wins the Chinese dark chess tournament. ICGA J. **42**(1), 53–56 (2020)
4. Davidson, A., Billings, D., Schaeffer, J., Szafron, D.: Improved opponent modeling in poker. In: International Conference on Artificial Intelligence, ICAI 2000, pp. 1467–1473 (2000)
5. He, H., Boyd-Graber, J., Kwok, K., Daumé III, H.: Opponent modeling in deep reinforcement learning. In: International Conference on Machine Learning, pp. 1804–1813. PMLR (2016)
6. Kim, M.-J., Kim, K.-J.: Opponent modeling based on action table for MCTS-based fighting game AI. In: 2017 IEEE Conference on Computational Intelligence and Games (CIG), pp. 178–180. IEEE (2017)
7. van den Herik, H.J., Donkers, H.H.L.M., Spronck, P.H.M.: Opponent modelling and commercial games. In: Proceedings of the IEEE Symposium on Computational Intelligence and Games, pp. 15–25 (2005)
8. Walczak, S.: Improving opening book performance through modeling of chess opponents. In: Proceedings of the 1996 ACM 24th Annual Conference on Computer Science, pp. 53–57 (1996)

Using Deep Learning to Detect Facial Markers of Complex Decision Making

Gianluca Guglielmo$^{(\boxtimes)}$ ⓘ, Irene Font Peradejordi ⓘ, and Michal Klincewicz ⓘ

Cognitive Science and Artificial Intelligence, Tilburg University, Warandelaan 2,
5037 AB Tilburg, The Netherlands
{g.guglielmo,m.w.klincewicz}@tilburguniversity.edu

Abstract. In this paper, we report on an experiment with The Walking Dead (TWD), which is a narrative-driven adventure game where players have to survive in a post-apocalyptic world filled with zombies. We used OpenFace software to extract action unit (AU) intensities of facial expressions characteristic of decision-making processes and then we implemented a simple convolution neural network (CNN) to see which AUs are predictive of decision-making. Our results provide evidence that the pre-decision variations in action units 17 (chin raiser), 23 (lip tightener), and 25 (lips part) are predictive of decision-making processes. Furthermore, when combined, their predictive power increased up to 0.81 accuracy on the test set; we offer speculations about why it is that these particular three AUs were found to be connected to decision-making. Our results also suggest that machine learning methods in combination with video games may be used to accurately and automatically identify complex decision-making processes using AU intensity alone. Finally, our study offers a new method to test specific hypotheses about the relationships between higher-order cognitive processes and behavior, which relies on both narrative video games and easily accessible software, like OpenFace.

Keywords: Video Games · Decision-Making · Facial Expression Machine Learning

1 Introduction and Related Work

1.1 Decision-Making in Video Games

Decision-making has been studied extensively in social psychology and economics with paradigms such as the prisoner dilemma, the ultimatum game, and the dictator game [3]. These paradigms are largely grounded in game theory, which

---

The research reported in this study is funded by the MasterMinds project, part of the RegionDeal Mid- and West-Brabant, and is co-funded by the Ministry of Economic Affairs, Region Hart van Brabant, REWIN, Region West-Brabant, Midpoint Brabant, Municipality of Breda and Municipality of Tilburg awarded to MML.

C. Browne et al. (Eds.): ACG 2021, LNCS 13262, pp. 187–196, 2022.
https://doi.org/10.1007/978-3-031-11488-5_17

assumes idealizations about rationality, utility, and often ignores the unique ways in which people make decisions in different contexts. Video games provide an alternative to game theory paradigms in the study of decision-making precisely because they provide a rich context for decisions in the form of a narrative, including in-game mechanics, and non-player characters (NPC) [25].

NPCs are important in moving video-game narratives forward and also in framing the decisions players make while playing. This framing typically involves consequences in the narrative of the game and expressions of emotions on the part of the NPCs. In this sense, decisions made in video games may involve similar cognitive and affective mechanisms that are at work during decision-making in real life, where meaningful decisions happen in a rich context with consequences that affect other people. The important difference, of course, is that consequences in video games affect the game world and NPCs, while decisions out-of-game affect the real world and real people. This difference, while a limitation, also makes video games useful in the study of complex decision-making, in that they provide a safe environment to experience new forms of agency without worries about the consequences [17]. This is also why video games are particularly useful in education [1]. Considering the aforementioned advantages, we decided to use TWD for our study, since its rich narrative presents scenarios that, to a certain extent, can be compared to the ones presented in real life.

1.2   Facial Expressions and Machine Learning

It is an old idea that the face is the window to the soul. Facial expressions have been systematically studied and linked to a set of basic emotions at least since Darwin [4], but have recently also been found to vary depending on the cultural context [13]. Emotions typically evoke a sympathetic system response. Being exposed to a stimulus, including making a decision, can also sometimes elicit a sympathetic response, which in turn changes heart rate, skin conductance, and facial temperature just as is the case with emotions [8,18]. Some of these responses, just as is the case with emotions, are accompanied by facial expressions. That said, not as much attention has been paid to the potential links between higher-order processes, such as decision-making, and facial expressions [9].

Facial expressions have been coded in the facial action coding system (FACS) developed by Paul Ekman and colleagues [7]. FACS is now used to measure pain in patients unable to communicate it verbally [15], and even in identifying depression [24]. Facial expressions are also widely used in affective computing, understood to be a research program that aims to use devices and systems to detect emotional states, processes, and responses [21].

Given all this, it is perhaps unsurprising that action units have been used as input for machine learning models. For example, a relatively simple support vector machine (SVM) reached 0.75 accuracy when using AUs as input for automatic stress detection [10]. SVM and k-nearest neighbors (KNN) algorithms can classify expressions of "pain" vs "no pain" and even their intensity [16,22]. More recently, CNNs have been used to estimate the presence of pain and its intensity [26]. In that last pain classification study, deep learning models had a higher accuracy when compared to other techniques; where the KNN algorithm implemented by [22] had an accuracy score of 0.86 and the CNN implemented by [26] had an accuracy score of 0.93. CNNs have also been used to detect emotions scoring an average beyond 0.92 on 8 classes of emotions [14]. AUs can also be combined with other input to further increase accuracy of a CNN model. Audio has been used with AUs for the detection of complex mental processes, such as depression [27] and to identify micro facial expressions [5]. Head and face rotation and the spatio-temporal dynamics occurring between AUs also increase accuracy of AU detection [19]. In sum, deep learning models, and in particular CNNs, are effective in detecting patterns in AUs to perform classification in different tasks. For this reason, we used them with AUs obtained during decision-making while playing TWD.

2   Methods

2.1   Data Collection and Participants

All participants were asked to play the first episode of TWD while seated in a room with another participant that did the same. All participants signed informed consent forms and were informed about the nature of the study and their rights regarding personal data storage and processing. Participants' gameplay was recorded using screen capture software and their posture and face were recorded using Open Broadcaster Software (OBS) and an HD Webcam (Logitech C922 Pro Stream); the two recordings were synchronized using a hotkey. The two participants taking part in any session of a recording always used two different computers, while the recordings were started and monitored using another two control computers.

A total of 78 participants took part in the experiment; 51 males with a mean age of 20.11 (SD = 2.63) and 27 females with a mean age of 19.4 (SD = 2.02). 12 participants were excluded since they played TWD before and knew the narrative and decisions presented in the game. One participant decided to quit the experiment because they found the content too disturbing. One participant had to leave due to personal issues and another 5 participants were excluded since they failed to perform the task as instructed. The final lot before data analysis had 52 participants. Game-play recordings were prepared with Sony Vegas Software by being cut into 10 s intervals around each decision made in the game. Each participant made 8 decisions during the experimental session, so a total of 80 s of video was eventually used to extract the information about AU intensity with OpenFace for each of the 52 participants.

2.2   Decision Selection

All of the decisions we used were important to the narrative of the game and relied on the participant taking into account the context in which they were presented by NPCs and the effect that their decision will have on the narrative of the game and NPCs (e.g., Fig. 1).

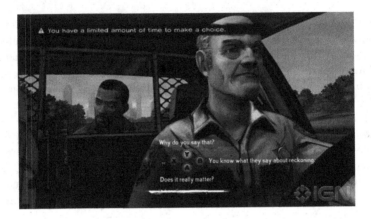

Fig. 1. An example of decision presented in TWD. The amount of time showed in a shrinking white bar on the lower part of the screen.

For example, in one of the decisions participants had to decide whether to save a young boy or an older man from zombies. While the consequences of these decisions would play out in the narrative of the game and affect NPCs, regardless of the decision made by the player, the video game followed a pre-defined course of action. So, each participant ultimately ended up playing the same section of the game with the same decisions. Importantly, the 8 decisions that were selected for analysis had more than 30 s between them. This eliminated the potential confound of effects of prior decisions overlapping with effects of the current decision.

3 Data Preparation and Modelling

3.1 Data Extraction

First, we identified the moment a decision was made by referencing the recording of game-play and the recording of the participant. We then used that moment as a representation of the end of the decision-making process and took 5 s of the video from before and 5 s after. For each of the 52 participants, eight 10 s videos were thus obtained, representing the 8 selected decisions made during TWD. The videos were recorded at 30 frames per second leading to a total of 300 frames, where the 150th frame represented the moment in which the decision was made. During this stage, we had to exclude a further 6 participants due to corrupted data or missing frames. Ultimately, 46 participants, with 8 videos each were used to extract AUs.

The AUs used for this work were extracted using OpenFace [2]. OpenFace extracts 17 action units (1, 2, 4, 5, 6, 7, 9, 10, 12, 14, 15, 17, 20, 23, 25, 26, and 45) that can be described either in terms of their presence (0 or 1) or in terms of their intensity (from 0 to 5). In our work, we extracted just the intensity information since, by itself, it can provide a number ranging from 0, the absence of the activity in the AUs, to 5, conveying the maximum intensity in the AUs. The data obtained were stored in CSV files.

3.2 Data Preprocessing

Since our focus was to detect facial AUs related to decision-making processes, we analyzed the 150 frames prior to the actual act of deciding corresponding to the click to finalize the decision. This is because we intended to focus on the processes prior to the decision itself. So, we compared the frames belonging to the baseline (0–74) to the frames belonging to the decision-making process (75–149). The 150 frames before making the decisions were equally split considering that the participants read the questions between frame 20 and 75 leaving frame 75–149 as the frames potentially reflecting the decision-making process. This particular split is motivated by the length of the sentences presented in the video game. Considering that the average speed to read 300 words per minute [23] and the eight sentences introducing the scenario had a number of words ranging from 4 to 10. Reading a 10-word sentence would require around 2 s, approximately corresponding to the 55 frames. For this reason, we considered frames 20 to 75 as a baseline period prior to the decision itself, which might have varied slightly according to the sentence length and the individual reader speed.

In the end, a total of 736 samples of AUs were used as input for the CNN: 46 participants had 8 recordings labelled as "baseline" and 8 recordings labelled as "decision-making process". Each of the 736 data points represented a row in the dataset. We then created a corresponding file with a 736×75 structure for each of the 17 AUs, where 736 is the number of total data points and 75 is the number of frames considered (representing the columns of the dataset), with half of the rows labeled "baseline" and half labeled "decision-making process". This allowed us to focus on each AU in isolation from others to examine its predictive power in classifying "decision-making" frames.

3.3 Model Description

In order to test the predictive value of individual AUs for identifying the decision-making processes, we created a 1D CNN, expecting it to serve as a baseline for more sophisticated modelling [28]. We decided to use CNNs since they have been successfully used with AUs for prediction and classification tasks [11], as mentioned in the introduction. Furthermore, CNNs were used to perform classification tasks using a dataset with fewer than 1000 data points, similarly to our own dataset [20]. In the end, our model had 2 convolutional layers, 2 max-pooling layers, and 4 fully connected layers; the structure of the model and its specification is illustrated in Fig. 2.

The activation function chosen was Rectifier Linear Unit (ReLU) as suggested in Gudi et al. [11]. The optimizer chosen for our CNN was the Nesterov-accelerated Adaptive Moment Estimation (Nadam) with a learning rate of 0.001.

Model: "sequential_12"

Layer (type)	Output Shape	Param #
conv1d_24 (Conv1D)	(None, 73, 100)	1000
dropout_48 (Dropout)	(None, 73, 100)	0
max_pooling1d_24 (MaxPooling	(None, 24, 100)	0
conv1d_25 (Conv1D)	(None, 22, 32)	9632
max_pooling1d_25 (MaxPooling	(None, 7, 32)	0
dropout_49 (Dropout)	(None, 7, 32)	0
flatten_12 (Flatten)	(None, 224)	0
dense_48 (Dense)	(None, 128)	28800
dropout_50 (Dropout)	(None, 128)	0
dense_49 (Dense)	(None, 180)	23220
dropout_51 (Dropout)	(None, 180)	0
dense_50 (Dense)	(None, 30)	5430
dense_51 (Dense)	(None, 2)	62

Total params: 68,144
Trainable params: 68,144
Non-trainable params: 0

Fig. 2. Model Specifications

In past studies, Nadam outperformed other optimizers in models that aimed to classify different typologies of data. More specifically, using Nadam resulted in lower convergence time required, lower loss score, and higher accuracy [6]. To minimize overfitting that might affect results on a small dataset like the one we used, dropouts were added between the convolutional, the max-pooling, and the fully connected layers. 20% of the AU dataset was used for test purposes, while 10% was used for validation and to keep track of potential overfitting. The model was trained using 10 sample mini-batches and 20 epochs. All of this was implemented in Python using Numpy, Pandas, Scikit-learn, and Keras libraries.

4   Results

The results suggest that three AUs might be predictive of decision-making processes. As shown in the Table 1 these units all scored above 0.65 (threshold used to select significant AUs).

The three action units make it possible to discriminate between decision-making processes and baseline even in a simple 1D CNN are: AU17 (chin raiser), AU23 (lip tightener), and AU25 (lips part). Other AUs did not reach significance

Table 1. Significant differences in action units across baseline and decisions

	Training		Validation		Test	
AU	Accuracy	Loss	Accuracy	Loss	Accuracy	Loss
17	0.6954	0.5842	0.7627	0.5338	0.7297	0.5279
23	0.6948	0.5854	0.6949	0.5576	0.6824	0.5397
25	0.7240	0.5277	0.6780	0.6214	0.7027	0.5735

with our model, so were excluded in the reported results, but a more sophisticated model may well find other AUs in the same areas of the face predictive. To further explore the predictive power of these 3 action units, we combined them in a multidimensional input (75,3) to the same network using the same number of epochs to obtain consistent results. The final model scored 0.81 on accuracy and 0.50 on loss score on the test set (Table 2).

Table 2. Combined significant AUs across baseline and decisions

Training		Validation		Test	
Accuracy	Loss	Accuracy	Loss	Accuracy	Loss
0.8144	0.4077	0.7966	0.4188	0.8108	0.4978

For a model this simple, our results suggest that AUs can indeed be used to identify decision-making processes without much modelling. To make it clear that this is not an anomaly, we include the convolution over epochs in Fig. 3 below.

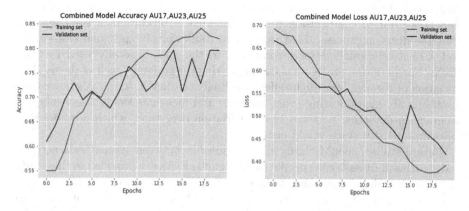

Fig. 3. Convolution of combined AUs 17, 23, and 25; 20 epochs (where 20, being the last number, is not displayed on the x-axis).

Our results seem to be corroborated by other studies. A study using AUs and SVM found that AU17, AU23, and AU25 intensities are modulated by stress conditions [10]. So, it might be the case that decision-making processes trigger a response similar to stressful stimuli [27]. Further corroborating our results, a distinct experimental study with TWD identified significant variations in temperature of the chin area approximately 20 s after decisions that had a moral dimension [12]. In other words, the same area that involves AU17, and AU25, shows a significant variation in temperature after particularly complex and possibly stressful decisions. The AU17 involves a tightening of the muscle mentalis, which is located below the lower lip, while AU25 involves the same muscle relaxation, so one explanation for distinctive variations of temperature in specific parts of the face is the effect of increased blood flow to those areas, which is caused by the engagement of muscles in facial regions [8], as identified in the present experiment with a CNN.

5   Discussion

Given their functional and anatomic connection, the predictive value of AU17 and AU23 might be a result of the tightening of the chin at the beginning of the decision-making process. AU25, on the other hand, is a functional counterbalance to AU17 and AU23 and logically connected to movements of the mouth and chin. Interestingly, AU26 (jaw drop), is functionally related to AU25, and while not included in the final model due to just-below 0.65 of accuracy, it seems to be engaged during the decision-making process as well. As a consequence, it might be the case that AU17 and AU23 are characteristic of the initial part of the decision-making process while AU25 and AU26 might be peculiar to the end part of the decision-making process when the facial expression returns to baseline. AU17 (chin raiser) seems to be counterbalanced by AU25 (lips part) while AU23 (lip tightener) might be counterbalanced by both AU25 and AU26 (jaw drop). In general, we can conclude that there is a tightening of the lip-chin area prior to the decision process and then a relaxation of the chin area after the decision process. That said, these results involve just one video game and a relatively small dataset compared to the ones generally used to train CNNs. Furthermore, processes occurring during the training (such as the random initiation of the weights) might affect the final accuracy in some AUs more than in others. So, incorporating other methods and measures would likely increase accuracy in detecting patterns in AU variation over time, which a model that relies on intensity alone would not.

Future studies should clarify the relationship between sympathetic activity and changes in intensity in specific facial regions. Evidence provided in this study suggests that decision-making is in some way connected to muscular activity in the chin area. This might in turn lead to changes in temperature, due to increased blood flow. These effects might be accentuated by stress caused by moral aspects that characterize some decisions. Moral decisions might be more stressful than non-moral ones thus eliciting a change in muscle activity and then in temperature. Therefore, future investigations should also pay special attention to the effects of decision-making and moral decision-making on AUs while keeping in mind the possible involvement (or confound) of stress on AU intensity. Ultimately, if it becomes possible to detect the moment of decision-making during game-play using the technique we outline here, our methods could prove useful in future game development.

References

1. Anastasiadis, T., Lampropoulos, G., Siakas, K.: Digital game-based learning and serious games in education. Int. J. Adv. Sci. Res. Eng. (ijasre) **4**(12), 139–144 (2018)
2. Baltrušaitis, T., Robinson, P., Morency, L.P.: Openface: an open source facial behavior analysis toolkit. In: 2016 IEEE Winter Conference on Applications of Computer Vision (WACV), pp. 1–10. IEEE (2016)
3. Conitzer, V., Sinnott-Armstrong, W., Borg, J.S., Deng, Y., Kramer, M.: Moral decision making frameworks for artificial intelligence. In: Proceedings of the AAAI Conference on Artificial Intelligence, vol. 31, pp. 4831–4835 (2017)
4. Darwin, C.: The Expression of the Emotions in Man and Animals. University of Chicago Press (2015). https://doi.org/10.7208/9780226220802https://doi.org/10.7208/9780226220802
5. Davison, A.K., Merghani, W., Yap, M.H.: Objective classes for micro-facial expression recognition. J. Imag. **4**(10), 119 (2018)
6. Dogo, E.M., Afolabi, O.J., Nwulu, N.I., Twala, B., Aigbavboa, C.O.: A comparative analysis of gradient descent-based optimization algorithms on convolutional neural networks. In: 2018 International Conference on Computational Techniques, Electronics and Mechanical Systems (CTEMS), pp. 92–99. IEEE (2018)
7. Ekman, P., Friesen, W.V.: Measuring facial movement. Environ. Psychol. Nonverbal Behav. **1**(1), 56–75 (1976)
8. Engert, V., Merla, A., Grant, J.A., Cardone, D., Tusche, A., Singer, T.: Exploring the use of thermal infrared imaging in human stress research. PloS one **9**(3), e90782 (2014). https://doi.org/10.1371/journal.pone.0090782[
9. Furl, N., Gallagher, S., Averbeck, B.B.: A selective emotional decision-making bias elicited by facial expressions. PLoS ONE **7**(3), e33461 (2012)
10. Giannakakis, G., Koujan, M.R., Roussos, A., Marias, K.: Automatic stress detection evaluating models of facial action units. In: 2020 15th IEEE International Conference on Automatic Face and Gesture Recognition (FG 2020), pp. 728–733. IEEE (2020)
11. Gudi, A., Tasli, H.E., Den Uyl, T.M., Maroulis, A.: Deep learning based facs action unit occurrence and intensity estimation. In: 2015 11th IEEE International Conference and Workshops on Automatic Face and Gesture Recognition (FG). vol. 6, pp. 1–5. IEEE (2015)
12. Guglielmo, G., Klincewicz, M.: The Temperature of morality: a behavioral study concerning the effect of moral decisions on facial thermal variations in video games. In: The 16th International Conference on the Foundations of Digital Games (FDG) 2021 (FDG'21). ACM, Motreal (2021). https://doi.org/10.1145/3472538.3472582
13. Jack, R.E., Garrod, O.G.B., Yu, H., Caldara, R., Schyns, P.G.: Facial expressions of emotion are not culturally universal. Proc. Natl Acad Sci. **109**(19), 7241–7244 (2012)

14. Liliana, D.Y.: Emotion recognition from facial expression using deep convolutional neural network. J. Phys. Conf. Ser. **1193**, p. 12004 (2019)

15. Lints-Martindale, A.C., Hadjistavropoulos, T., Barber, B., Gibson, S.J.: A psychophysical investigation of the facial action coding system as an index of pain variability among older adults with and without Alzheimer's disease. Pain Med. **8**(8), 678–689 (2007)

16. Lucey, P., Cohn, J.F., Prkachin, K.M., Solomon, P.E., Chew, S., Matthews, I.: Painful monitoring: automatic pain monitoring using the UNBC-McMaster shoulder pain expression archive database. Image Vis. Comput. **30**(3), 197–205 (2012)

17. Nguyen, C.T.: Games and the art of agency. Philos. Rev. **128**(4), 423–462 (2019)

18. Ohira, H., et al.: Neural mechanisms mediating association of sympathetic activity and exploration in decision-making. Neuroscience **246**, 362–374 (2013)

19. Onal Ertugrul, I., Yang, L., Jeni, L.A., Cohn, J.F.: D-PAttNet: Dynamic patch-attentive deep network for action unit detection. Front. Computer Sci. **1**, 11 (2019)

20. Pasupa, K., Sunhem, W.: A comparison between shallow and deep architecture classifiers on small dataset. In: 2016 8th International Conference on Information Technology and Electrical Engineering (ICITEE), pp. 1–6 (2016)

21. Picard, R.W., Picard, R.: Affective Computing, vol. 252. MIT Press, Cambridge (1997)

22. Prkachin, K.M., Solomon, P.E.: The structure, reliability and validity of pain expression: evidence from patients with shoulder pain. Pain **139**(2), 267–274 (2008)

23. Rayner, K., Schotter, E.R., Masson, M.E.J., Potter, M.C., Treiman, R.: So much to read, so little time: how do we read, and can speed reading help? Psychol. Sci. Public Interest **17**(1), 4–34 (2016). https://doi.org/10.1177/1529100615623267

24. Reed, L.I., Sayette, M.A., Cohn, J.F.: Impact of depression on response to comedy: A dynamic facial coding analysis. J. Abnorm. Psychol. **116**(4), 804 (2007)

25. Ryan, M., Formosa, P., Howarth, S., Staines, D.: Measuring morality in videogames research. Ethics Inf. Technol. **22**(1), 55–58 (2019). https://doi.org/10.1007/s10676-019-09515-0

26. Semwal, A., Londhe, N.D.: automated pain severity detection using convolutional neural network. In: 2018 International Conference on Computational Techniques, Electronics and Mechanical Systems (CTEMS), pp. 66–70. IEEE (2018)

27. Song, S., Shen, L., Valstar, M.: Human behaviour-based automatic depression analysis using hand-crafted statistics and deep learned spectral features. In: 2018 13th IEEE International Conference on Automatic Face & Gesture Recognition (FG 2018), pp. 158–165. IEEE (2018)

28. Tang, W., Long, G., Liu, L., Zhou, T., Jiang, J., Blumenstein, M.: Rethinking 1d-cnn for time series classification: A stronger baseline. arXiv preprint arXiv:2002.10061 (2020)

A Generic Approach for Player Modeling Using Event-Trait Mapping Supported by PCA

M. Akif Gunes$^{(\boxtimes)}$ ⓘ, M. Fatih Kavum$^{(\boxtimes)}$ ⓘ, and Sanem Sariel$^{(\boxtimes)}$ ⓘ

Artificial Intelligence and Robotics Laboratory, Faculty of Computer and Informatics Engineering, Istanbul Technical University, Istanbul, Turkey
{gunesme,kavum,sariel}@itu.edu.tr
https://air.cs.itu.edu.tr/

Abstract. Modeling players based on their in-game events is essential for predicting their future behaviors. Player modeling studies mostly target a specific game or genre. This makes it difficult to transfer existing methods from one game to another. In this study, we propose a generic event-trait mapping and unsupervised learning approach for player modeling that extends our earlier modeling method with Principal Component Analysis (PCA). We present a case study of this new approach on a dataset of ten thousand players of World of Warcraft (WoW), a massive multiplayer online role-playing game (MMORPG). The base and the extended approaches are compared with an AutoEncoder (AE) based approach on this dataset. The methods generate clusters (persona) as mixtures of different character traits. The best results are obtained with the extended event-trait mapping approach for player modeling.

Keywords: Player Modeling · Player Profiling · Event-Trait Mapping · PCA · World Of Warcraft

1 Introduction

The process of personality prediction in video games is referred to as player modeling. Player modeling not only helps with personality prediction but is also an imperative tool in the game industry for monetary gain. It is key to adjust, improve, and develop products for different types of players to improve their satisfaction. The first and the most crucial step is to model players in a fast and reliable way either in real-time or during development. Player modeling approaches may vary for different games since one may wish to emphasize a distinctive trait of the game for a better result. Therefore, modeling different games with a common method is a challenge. Some generic models [18] use common aspects of games of different genres. They generalize game actions (events) to group them into smaller chunks and apply profiling for different genres, showing that a generic approach is plausible. However, this approach is not applicable for games with no common ground (no common group of events).

© Springer Nature Switzerland AG 2022
C. Browne et al. (Eds.): ACG 2021, LNCS 13262, pp. 197–207, 2022.
https://doi.org/10.1007/978-3-031-11488-5_18

In this study, we propose a generic unsupervised learning method that can be applied to a variety of games genres. Our method builds on our earlier work, Event-Trait Mapping and Feature Weighting method (ET-FW) [12] and extends it (ET-PCA) with the use of Principal Component Analysis (PCA) [15]. ET-FW creates profiles as combinations of one or more personality traits in different proportions. Associating events with traits not only provides a generic approach but also higher quality and interpretable cluster distribution compared to that of clustering by only using events.

We present a case study of ET-PCA on the World of Warcraft (WoW) [19], a massive multiplayer online role-playing game (MMORPG) that contains a wide spectrum of different in-game elements found in almost every game genre [8]. We compare the results with the base method and an AutoEncoder (AE) based learning method.

The rest of this paper is organized as follows. Section 2 presents a review of literature. The methodology is described in Sect. 3. The case study on WoW data is presented in Sect. 4 which also includes the experimental results. Finally, concluding remarks are made in Sect. 5.

2 Related Works

There is a rich body of literature on player profiling/modeling. We present the ones targeting most relevant ones for our study. Halim et al. [13] explore the player profiles with three different feature selection algorithms, four different clustering techniques and three different classifier methods on three different games. They use a personality test on users to label them and select the traits as openness, conscientiousness, extraversion, agreeableness, and neuroticism. In this work, manual labeling of players is needed by giving them personality tests in real life.

Yee et al. propose a trait list and pre-processed event data for WoW [19]. Their study includes forty-six events and the necessary calculation methods. They investigate events which can provide insights about a person's personality. They also analyze the possibility of learning a person's personality just by studying their virtual behaviors. We inspire from some of these pre-processing steps that are used for WoW events. Brown and Mitchell used a questionnaire to measure personality tests based on five factors while measuring the relationship between personality and gambling style over poker games [16]. Wang et al. [5] present a behavior prediction method by using an auto-encoding neural network but not on WoW data. They use traditional AE method as a pre-training step. Charles and Cowley use archetypal profiles other than trait profiles for their behavlet analytics-based profiling method [7]. They also use domain-expert knowledge for creating the behavlets. Different from earlier work, we propose a generic Event-Trait mapping method that can model and cluster players. Our approach does not rely on manual labeling.

3 Methodology

The main goal in this research is to determine in-game personality traits from players' in-game events. Player behaviors are represented via traits, which can be treated as classes. Players' trait scores are modeled using in-game events as features. Our work builds on our previous work, Event-Trait Mapping and Feature Weighting method (ET-FW) [12] which has shown to be successful in clustering players to traits for a casual mobile racing game called Dusk Racer. The selected traits for this game are explorer, meticulist, competitor, compulsivist, strategist, hoarder, social, exploiter. The first step of this clustering process is to determine event-trait relations. For this purpose, event-weight (EW), event-trait (ET), trait-event (TE) and user-event (UE) matrices are constructed to produce the final user-trait (UT) matrix.

- *EW*: represents the weights of the preprocessed events. It is a diagonal square matrix, with the size of event × event. The weights can be either determined by a domain expert or generated with the extracted data. In this research, frequencies of events has been used for weighting.
- *ET*: is a mapping matrix between in-game events and traits. This relation is determined by a domain expert. The scores are given from 0 to 5 where 0 means *not related* and 5 means *fully related*. An event can be mapped to more than one trait. As stated in the review study by Hooshyar et al. [14], mapping of in-game events to traits is carried out by experts in the majority of profiling processes. We also prefer this way when doing mapping.
- *TE*: Just like the event-trait matrix, this matrix is filled with the same intuition. This time, traits are mapped to events. This relation is determined by a domain expert. The size and dimensions of ET and TE are the same. Again, the scores are given from 0 to 5.
- *ET ∘ TE*: Represents the final mapping. Note that the multiplication operation here is an element-wise operation called Hadamard product [9]. This gives a resulting matrix of the same size of ET and TE whose elements take on values from 0 and 25.
- *UE*: Represents the frequencies of events done by each user. Since the total number is different for each event, it is normalized, used and obtained automatically.
- *UT*: This matrix shows the relation between users and traits. It is the final product, and computed by Eq. 1. Normalization is done using the min-max scaling.

$$UT = normalize(UE \times EW) \times (ET \circ TE) \tag{1}$$

After the computation of UT, the Expectation-Maximization (EM) method is applied to cluster users with this matrix. To see each persona trait ratio, mean trait values given by the EM method for each cluster is used. The details of these computations are presented in our previous work [12].

Mapping traits to game events reveal different behavior groups of the players and their distributions, leading to a better understanding of the player-game interaction. In order to better distinguish these player groups, we further extend our previous work with Principal Component Analysis (PCA). The new method, Event-Trait Mapping extended with PCA (ET-PCA), improves the cluster quality and interpretability with the use of PCA. PCA is an automated exploratory data analysis process to find principle components in a dataset which can also be used for dimensionality reduction [15].

In traditional profiling methods, the game logs are reduced to a certain number of records using PCA, and then the players can be profiled by any clustering method [11], yet in complex games, these clusters are very difficult to interpret. Furthermore, since PCA is an unsupervised learning method, it determines dimensions of maximum variance without reference to class labels. To apply PCA, the sum-product of the trait and variance values of events in each PCA component is taken to create PCA component/Trait relation (Eq. 2).

$$sp = \sum_{i=0}^{n} a_i * b_i \tag{2}$$

where sp is the sumProduct value, n is the index of events in the PCA component, a_n is the variance of the particular event in that PCA component, and b_n is the trait value taken from the ET matrix. Then, k-means clustering algorithm [4] is applied to the selected first-6 PCA components to create clusters. Finally, cluster trait relations are obtained using this relation as the result of the ET-FW method by applying the sum product operation between "PCA component/Trait" matrix and "PCA component/mean values of clusters" relations.

The following section presents a case study of this method on the WoW game. WoW is picked since it includes a wide spectrum of different in-game elements found in almost every game genre [8].

4   A Case Study on Wow Data

4.1   Data Collection and Preprocessing

Player data used in this study are collected from a third party website, WoW-Progress [1]. More than 650000 guild informations can be accessed here. Guilds are formed by players for better player experience and represent in-game association of player characters in the game. Players collaborate to form teams, and they socialize and help each other in their guild. These guild data are categorized by the WoWProgress company according to language, realm and tier in their website. Blizzard, the original publisher of the World of Warcraft game, has also set up an API system to the benefit of the developers [2].

In order to access the detailed game logs of these players, 51 different API functions are called for each player, and more than 10000 events are obtained. The events which are not directly correlated to the traits in the raw data are eliminated, and the remaining events are preprocessed, yielding 175 events. In the preprocessing phase, some of the stats are rewritten as averaged percentages, or rates[1]. Because it is known that the player levels which is a measure of character

---

[1] Event explanations are available at: https://playerprofiling.github.io/WoWEvents/.

progress fluctuates greatly in terms of events, only 120 level players are taken into consideration for our experiments. Eventually, 11958 players are registered in the database with 175 events.

4.2 Trait Selection

The selected traits in our study are mostly inspired from Bartle's [6] and Ferro et al.'s studies [10]. According to Bartle's study, personas are considered to have four main traits, namely; killers, achievers, socializers and explorers. Ferro et al. extend the edition of the Bartle's Player Type Graph. Based on these studies and domain expert feedback, the traits are selected as follows: competitive (compete with others), casual (do not think too much about choices), explorer (explore the universe of the game), grinder (perform repetitive tasks), social (establish long-term relationships with other), craftsman (keep professions in the game ahead of the other traits), supportive (help others), and DPS-player (prone to kill all other players and NPCs)[2].

After the traits are selected, Event-Trait mapping is performed. The mapping process is done by taking average scores of four experts playing the game[3].

4.3 Applying Profiling Steps

ET-FW is applied on the ET and UE matrices. The resulting stacked graph for the trait clusters can be seen in Fig. 1. The results show that there are 8 different clusters with 8 different trait ratios. The top three highest-valued traits are considered for cluster representation. For example, the first cluster in Fig. 1 involves the grinder trait as the most dominant trait among the others. DPS-player and Crafter traits follow this trait for that cluster.

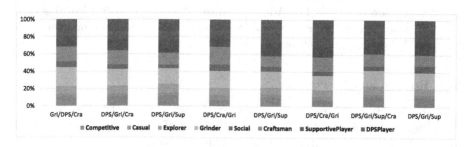

Fig. 1. The resulting stacked graph of the Trait-Cluster relation for ET-FW.

Then, our new method, ET-PCA, is applied on the same ET and UE matrices. After that, a new matrix that shows the relationship between PCA components and clusters is created. Then, k-means algorithm is applied to find the

---

[2] The details on the trait explanations are available at: https://playerprofiling.github.io/WoWTraits/.

[3] A sample Event-Trait matrix @see: https://playerprofiling.github.io/EventTrait/.

PCA-Cluster matrix. Considering that there may be a cluster for each trait, k is determined as 8. The sum product formula given in Eq. 2 is used between these two relations to create the final stacked graph (Fig. 2).

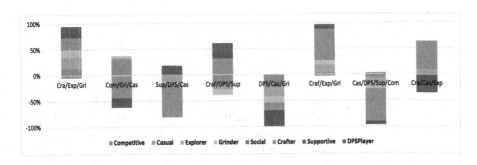

Fig. 2. Stacked Graph of Trait-Cluster relation for ET-PCA

For comparing our result with a deep neural network approach, the same AE approach used in Wang et al.'s work [5] is used within our study, except some of the parameters and the activation function are changed to fit our data needs. The AE method is applied to the WoW dataset and 175 preprocessed events are reduced to eight encoded events. Creating a Trait/Cluster relation similar to the other methods is not possible since the AE method reconstructs the input nodes in the output and there are no interpretable relationships between the original nodes and the encoded ones. For this reason, at the end of the profiling process, the players are clustered without having any information about their mixing rate of character traits, but only their clusters. Leaky Relu [17] is used as the activation function. The Adam optimizer [20] with a 0.0005 learning rate is used. Mean Square Error is used as the Loss function, and the program was run with 200 epochs and 256 batch sizes.

4.4 Results and Discussion

First, the personas (cluster results) are compared between the three methods according to the selected traits. Then, the quality of the clusters are compared by analyzing their trait ratios and using silhouette analysis.

Evaluating Methods with Two Different Personas. In order to better explain this section, cluster IDs are assigned to the outcomes of the profiling methods. For ET-FW, ET-PCA and AE methods these clusterIDs are simple index numbers starting from 0 to 7. Also, they are in the same order as the clusters in the stacked chats given in Fig. 1 and Fig. 2.

Table 1. Top 15 competitor players and their assigned clusters for all methods. The order of displaying the competitor character trait relative to the other clusters is shown as rank. We can't report the rank for the AE method since it reconstructs the input nodes in its output. So it is not possible to have a proper relationship between original and encoded nodes.

UserId	Competitor Trait				
	ET-FW		ET-PCA		AE
	Cluster ID	Competitor Rank	Cluster ID	Competitor Rank	Cluster ID
778	2	1	1	1	2
1448	2	1	0	2	2
888	2	1	0	2	2
4786	2	1	0	2	2
473	6	2	1	1	2
3873	1	3	1	1	4
5433	6	2	1	1	1
6273	6	2	0	2	2
2008	1	3	1	1	4
5240	6	2	0	2	5
293	2	1	1	1	2
114	6	2	1	1	4
1425	6	2	1	1	4
5079	6	2	0	2	2
6800	1	3	1	1	3

Firstly, top fifteen players that have competitive character trait more than other ten-thousand players are selected by taking the players with the highest number of "ArenasPlayed", "DuelsWon" and "WorldHonorableKills" events.

Table 1 reports the results for these players.

The top three clusters which have the highest competitive trait values for the ET-FW method are Cluster 2 (trait value: 130.53), Cluster 6 (trait value: 102.38) and Cluster 1 (trait value: 80.13). These trait values for the ET-FW method are taken from the Trait-Cluster matrix that is obtained previously to create ET-FW stacked graph (Fig. 1). The order of displaying the competitor character trait relative to the other clusters is shown as rank in Table 1. Hence, the competitor ranks for these clusters are 1, 2 and 3, respectively. The distribution is the same in Table 1 for the ET-FW method. So, for the competitive trait, the ET-FW method puts the selected players into the most related three clusters.

For the ET-PCA method, the top three clusters that have the highest competitive trait values are Cluster 1 (trait value: 33.63), Cluster 0 (trait value: 14.92), and Cluster 5 (trait value: 8.5). Trait values are taken from the Trait-Cluster matrix that is obtained previously to create ET-PCA stacked graph (Fig. 2).

Table 2. Top 15 grinder players and their assigned clusters for each method. The order of displaying the character trait relative to other clusters shown as rank. We can't report the rank in AE due to its constraints.

UserId	Grinder Trait				
	ET-FW		ET-PCA		AE
	Cluster ID	Grinder Rank	Cluster ID	Grinder Rank	Cluster ID
9132	2	1	5	1	2
425	2	1	5	1	0
7069	2	1	5	1	0
3764	2	1	0	2	2
5423	6	2	5	1	0
6850	2	1	5	1	2
225	2	1	5	1	2
1448	2	1	0	2	2
480	2	1	5	1	2
3886	2	1	5	1	2
901	2	1	5	1	6
425	2	1	5	1	0
5423	6	2	5	1	0
7909	6	2	5	1	5
7076	2	1	5	1	2

These three clusters are the only clusters that have the positive value for the competitor trait. Thus, it can be said that the other clusters are not related to the competitor trait. Non-related cluster information cannot be provided, since there is no negative value for traits. By observing Table 1, it can be said that ET-PCA offers slightly better results while selecting players' clusters since it does not generate rank-3 clusters as ET-FW does. To sum up, while 3 different clusters are observed on the ET-FW side, there are 2 clusters corresponding to the top related traits for ET-PCA.

For AE, eight out of fifteen players are clustered in the same group (Cluster 2), and others are distributed separately in Clusters 4, 1, 5 and 3. A cluster/trait relationship cannot be obtained for AE; therefore, this method cannot be evaluated like the other two.

Users with grinder personas are selected by taking the players that have the highest number of "EpicItemsLooted" and "NeedRollsMadeOnLoot" events from the remaining ten-thousand people. Table 2 reports the results for these players.

The top three clusters that have the highest grinder trait values for ET-FW method are Cluster 2 (trait value: 190.62), Cluster 6 (trait value: 154.4) and

Cluster 1 (trait value: 134.00). From Table 2, it might be seen that 12/15 of the selected players are in the first grinder rank cluster.

For the ET-PCA method, the top three clusters with the highest grinder trait values are Cluster 5 (trait value: 22.07), Cluster 0 (trait value: 17.15) and Cluster 1 (trait value: 5.01). In addition, as can be seen in Fig. 2, there is also Cluster 7 which has the positive number for the grinder trait and the rest of the traits have negative values. From Table 2, it might be seen that 13/15 of the selected players are in the first grinder rank cluster.

For AE, eight out of fifteen players are in Cluster 2, and the other players are in Cluster 0 and 5. Yet, the quality of clusters with respect to traits cannot be assessed. More results are shown in our website[4].

Table 3. Mean Silhouette Coefficient analysis results for ET-FW and ET-PCA.

Cluster	MSC (ET-FW)	MSC (ET-PCA)
cluster0	0.13	0.18
cluster1	−0.11	0.14
cluster2	0.19	0.10
cluster3	0.18	0.10
cluster4	0.07	0.16
cluster5	0.19	0.18
cluster6	0.06	0.16
cluster7	−0.05	0.19
Overall	0.05	0.16

Comparing Cluster Quality. Silhouette analysis is applied to the generated player clusters for statistical analysis. The Silhouette Coefficient is a useful metric for evaluating clustering performance [3] and computed using Eq. 3:

$$sc = \frac{b - a}{max(a, b)} \tag{3}$$

where a is the mean intra-cluster distance, and b is the mean inter-cluster distance to the closest cluster. The score ranges from −1.0 to 1.0, where higher the score, the better is the outcome.

The results reported in Table 3 show that, ET-PCA has a better overall clustering performance than that of ET-FW.

5   Conclusion

In this paper, we present an Event-Trait Mapping based method extended with the use of PCA (ET-PCA) and a case study of this method on the WoW data.

---

[4] For more result tables: @see: https://playerprofiling.github.io/ResultTables/.

Persona clusters are created with this method as combinations and mixtures of different character traits. Two personas are selected for evaluating the resulting clusters on the WoW data. To this end, the selected top ten players' clusters are analyzed. The results indicate that ET-PCA groups players to the most related clusters. In addition, with this new method, traits can take on negative values. Having negative trait values, a good indicator, in a cluster means that players do not showcase those particular trait features. Furthermore, when the trait ratios of the clusters are observed on a stacked graph, it is seen that more visualizable results are obtained by the ET-PCA modeling method.

For future work, automated methods are planned to be developed for creating Event-Trait matrix. In the short term, a validation procedure is planned to correct human errors in creating matrices.

References

1. https://www.wowprogress.com/
2. https://develop.battle.net/
3. Rousseeuw, P.J.: Silhouettes: a graphical aid to the interpretation and validation of cluster analysis. J. Comput. Appl. Math. **20**, 53–65 (1987)
4. Ahmad, A., Dey, L.: A k-mean clustering algorithm for mixed numeric and categorical data. Data Knowl. Eng. **63**(2), 503–527 (2007)
5. Wang, S., Wang, H., Gao, Q., Hao, L.: Auto-encoder neural network based prediction of Texas poker opponent's behavior. Entertain. Comput. **40**, 100446 (2022)
6. Bartle, R.: Hearts, clubs, diamonds, spades: Players who suit muds. J. Virtual Environ. **1**(1), 19 (1996)
7. Charles, D., Cowley, B.U.: Behavlet analytics for player profiling and churn prediction. In: Stephanidis, C., et al. (eds.) HCII 2020. LNCS, vol. 12425, pp. 631–643. Springer, Cham (2020). https://doi.org/10.1007/978-3-030-60128-7_46
8. De Simone, L., Gadia, D., Maggiorini, D., Ripamonti, L.A.: Design of a recommender system for video games based on in-game player profiling and activities. Association for Computing Machinery, New York (2021)
9. Elrawy, A., Kishka, Z., Saleem, M., Abul-Dahab, M.: On hadamard and kronecker products over matrix of matrices. General Lett. Math. **4**, 13–22 (2018)
10. Ferro, L.S., Walz, S.P., Greuter, S.: Towards personalised, gamified systems: an investigation into game design, personality and player typologies. Association for Computing Machinery, New York (2013)
11. Gow, J., Baumgarten, R., Cairns, P., Colton, S., Miller, P.: Unsupervised modeling of player style with LDA. IEEE Trans. Comput. Intell. AI Games **4**(3), 152–166 (2012)
12. Gunes, M.A., Solak, G., Akin, U., Erden, O., Sariel, S.: A generic approach for player modeling using event-trait mapping and feature weighting. In: Proceedings of the AAAI Conference on Artificial Intelligence and Interactive Digital Entertainment, vol. 12, pp. 169–175 (2016)
13. Halim, Z., Atif, M., Rashid, A., Edwin, C.A.: Profiling players using real-world datasets: clustering the data and correlating the results with the big-five personality traits. IEEE Trans. Affect. Comput. **10**(4), 568–584 (2019)
14. Hooshyar, D., Yousefi, M., Lim, H.: Data-driven approaches to game player modeling: a systematic literature review. ACM Comput. Surv. **50**(6), 1–19 (2018)

15. Jolliffe, I.: Principal component analysis. In: Lovric, M. (ed.) International Encyclopedia of Statistical Science, pp. 1094–1096. Springer, Heidelberg (2011). https://doi.org/10.1007/978-3-642-04898-2_455

16. Brown, S.C., Mitchell, L.A.: An observational investigation of poker style and the five-factor personality model. J. Gambling Stud. **26**, 229–34 (2009). https://doi.org/10.1007/s10899-009-9161-9

17. Li, Y., Yuan, Y.: Convergence analysis of two-layer neural networks with ReLU activation. In: Proceedings of the 31st International Conference on Neural Information Processing Systems, NIPS 2017, pp. 597–607. Curran Associates Inc., Red Hook (2017)

18. Shaker, M., Shaker, N., Togelius, J., Abou-Zleikha, M.: A progressive approach to content generation. In: Mora, A.M., Squillero, G. (eds.) EvoApplications 2015. LNCS, vol. 9028, pp. 381–393. Springer, Cham (2015). https://doi.org/10.1007/978-3-319-16549-3_31

19. Yee, N., Ducheneaut, N., Nelson, L., Likarish, P.: Introverted elves & conscientious gnomes: the expression of personality in world of warcraft, vol. 2011, pp. 753–762 (2011)

20. Zhang, Z.: Improved adam optimizer for deep neural networks. In: 2018 IEEE/ACM 26th International Symposium on Quality of Service (IWQoS), pp. 1–2 (2018)

Game Systems

Automatic Generation of Board Game Manuals

Matthew Stephenson[✉], Éric Piette, Dennis J. N. J. Soemers,
and Cameron Browne

Department of Data Science and Knowledge Engineering, Maastricht University,
Paul-Henri Spaaklaan 1, 6229 EN Maastricht, The Netherlands
{matthew.stephenson,eric.piette,dennis.soemers,
cameron.browne}@maastrichtuniversity.nl

Abstract. In this paper we present a process for automatically generating manuals for board games within the Ludii general game system. This process requires many different sub-tasks to be addressed, such as English translation of Ludii game descriptions, move visualisation, highlighting winning moves, strategy explanation, among others. These aspects are then combined to create a full manual for any given game. This manual is intended to provide a more intuitive explanation of a game's rules and mechanics, particularly for players who are less familiar with the Ludii game description language and grammar.

Keywords: Ludii · Board Games · Manuals · Tutorial · Procedural Content Generation

1 Introduction

Board games are one of the most popular pastimes for millions of people, and have been played for over 5000 years [3]. Board Game Geek, one of the most popular repositories of board games, currently includes over 130,000 different games.[1] As such, manually describing the rules for all these games in a clear and consistent manner is a near impossible task. In this paper we present a game manual generation framework, which automatically creates explanations for how to play any given board game. This framework operates on the Ludii general game system, which supports the majority of non-dexterity board games.

The rest of this paper is organised as follows. Section 2 provides the necessary background information about Ludii and other related work. Section 3 describes the Ludii game description language and our process for translating it into English. Section 4 describes how different moves for a given game are detected, classified and visualised. Section 5 describes several additional features that can provide supplementary information. Section 6 describes how these previous aspects are combined into a complete game manual. Section 7 concludes this work, discusses several limitations, and suggests ideas for future research.

---

[1] https://boardgamegeek.com.

© Springer Nature Switzerland AG 2022
C. Browne et al. (Eds.): ACG 2021, LNCS 13262, pp. 211–222, 2022.
https://doi.org/10.1007/978-3-031-11488-5_19

2 Background

2.1 Ludii

Ludii is a general game system [17] that is being developed as part of the Digital Ludeme Project [3]. The majority of games within Ludii are traditional board games, but other types of games such as puzzles, finger games, dice games, etc., are also present. Ludii is also able to support games which are stochastic or contain hidden information. Ludii currently includes over 950 playable games,[2] covering a wide-range of different categories and mechanics.

2.2 Related Work

Video Games. The work that most closely resembles our desired outcome would probably be the AtDelfi system [7] for the General Video Game AI (GVGAI) framework [14,15]. This system generates instruction cards for simple arcade-style video games written in the video game description language (VGDL), providing information about the game's controls, how to gain points, and how the game is won or lost. This information is displayed using a combination of text, images, and GIF animations. Although this approach focuses on video games rather than board games, the generated instruction cards are similar in design and purpose to the manuals we would like to produce.

However, there are several additional considerations when creating manuals for board games. Board games often have very different control schemes to those of video games, where moves are defined less by the exact buttons the player can press and more by what pieces can be placed or moved in accordance with the game's rules. Video games also typically have specific considerations which do not apply to board games, such as non-player characters (NPCs), projectiles, timed events, a larger state space, etc. Likewise, board games often contain aspects such as multiple players, distinct game-play phases, a greater reliance on strategy rather than dexterity, a larger action space, etc. Because of this, our generated board game manuals will likely contain very different information from AtDelfi, with a much greater emphasis on piece movement and the specific rules of play.

Learning from Observation. This work involves learning the rules of board games from observed play [2,6,9,13], where a learner agent is given a collection of play traces and tasked with learning the rules of the game which produced them. Prior work in this area has predominately focused on learning from games written in the Stanford Game Description Language (GDL), which is a much more verbose and low-level language compared to that of Ludii. This makes direct translation from the game description very difficult, hence the reliance on learning from observed moves. Research on this task is also somewhat limited,

---

[2] https://ludii.games/download, version 1.2.9.

```
(game "Tic-Tac-Toe"
   (players 2)
   (equipment {
      (board (square 3))
      (piece "Disc" P1)
      (piece "Cross" P2)
   })
   (rules
      (play (move Add (to (sites Empty))))
      (end (if (is Line 3) (result Mover Win)))
   )
)
```

Fig. 1. Game description for Tic-Tac-Toe, written in Ludii's game description language.

with presented approaches so far only being able to translate a limited subset of GDL game descriptions.

Our presented approach instead relies more heavily on the higher level language provided by Ludii, utilising the ability to specify English translations for specific sub-sections of a games's description based on its wider context. This arguably makes our approach less general than learning from observation, as it is currently unable to operate on games outside of Ludii. However, our approach works effectively on the majority of board games within Ludii, and the presented ideas are likely to be applicable to other high-level game description languages with only minor modifications.

3   Ludii Game Description Language

Every game within Ludii is described as a single symbolic expression, which contains structured sets of pre-defined keywords and values. These keywords are called ludemes, and are intended to represent some fundamental aspect of a game (e.g. Line, Mover, Win). Some ludemes are atomic and require no additional parameters, such as the examples given above, while others such as if require additional arguments to be provided, in this case a condition and statement. By combining several ludemes and values together into compound expressions, we can create larger ludemeplexes that describe more complex ideas. For example, (if (is Line 3) (result Mover Win)) describes that forming a line of 3 pieces results in a win. This same idea is used to describe all of the games in Ludii, repeatedly creating and combining increasingly complex ludemeplexes until the final game description is obtained. As an example, the complete Ludii game description for Tic-Tac-Toe is shown in Fig. 1.

The number of ludemes needed to describe a game varies based on its complexity, typically ranging from only a few dozen for simple examples like Tic-Tac-Toe and Hex, to several hundred for more complex cases like Chess and Backgammon. The complete set of ludemes within Ludii is referred to as the Ludii Game Description Language (L-GDL), and is automatically inferred from

The game "Tic-Tac-Toe" is played by two players on a 3x3 rectangle board
    with square tiling.
Player one plays with Discs. Player two plays with Crosses.
Players take turns moving.
Rules:
    Add one of your pieces to the set of empty cells.
Aim:
    If a player places 3 of their pieces in an adjacent direction line,
        the moving player wins.

Fig. 2. Automatic English language translation of Ludii's Tic-Tac-Toe game description.

the Ludii codebase using a class grammar approach [5]. Further details on L-GDL are provided in the Ludii game logic guide [16].

3.1   Translation to English

While Ludii game descriptions are generally clear and understandable for simple cases like Tic-Tac-Toe, deciphering more complex game descriptions often requires expert knowledge about L-GDL. As a result, Ludii relies on hand-written descriptions to accompany each game. These are often obtained from Board Game Geek, Wikipedia, or other sources with inconsistent levels of detail and structure. Even worse, some of these descriptions might make allusions to other games which players may be unfamiliar with, e.g. "Knights move the same as in Chess". All these issues motivate the creation of an automated game-translation process, which is able to convert any Ludii game description into a pseudo-English equivalent.

Due to the class grammar approach used to generate L-GDL, each Ludeme has a corresponding class within the Ludii codebase. Within each of these classes we define a toEnglish() function, which returns an English language description of how the ludeme operates within the game. These descriptions often need to consider the ludeme's arguments, for example (is Line 3) might translate to "3 pieces in a line". As such, the toEnglish() function for each ludeme will often need to call the toEnglish() functions of its arguments. For example, the toEnglish() function of the if ludeme also calls the toEnglish() of its conditional and statement arguments. This structure means that calling the toEnglish() function on the highest level (game ...) ludeme of any given game description, will recursively call and combine the results of all its sub-ludemes to create a full English language translation of the game. As an example, applying this process on the description in Fig. 1, outputs the text shown in Fig. 2. Additional examples of English translations of Ludii game descriptions are shown in Appendix A.

However, due to the complexities and inconsistencies of the English language, this translation process is unlikely to work perfectly for every case. This includes mostly harmless grammar issues like displaying the correct pluralisation of different words, to more complex challenges like converting a sequence of nested statements such as (or A (or (or B C) D) (or E F)) into an unambiguous sentence.

Ludii also contains certain implementation-specific ludemes such as *SetPending* or *SetState*, which are often used for different purposes across multiple games. For example, in the game Chess the state of a piece indicates whether it has previously been moved or not, and is used to determine if castling is possible. Alternatively, in the game Jungle the state of a piece indicates its combat strength, and is used to determine if certain pieces can capture others. As of the time of writing, our translation approach is currently unable to directly link the action of setting a piece's state with the possible future consequences of this.

4 Move Visualisations

While the English translation of a game's description provides an explanation of the possible moves a player can make, there are some moves which are much easier to understand visually. Chess provides a nice example of this, where showing the movements of the different pieces such as Knight, Queen, Rook, and Bishop using diagrams or animations may be much more intuitive than with text alone. Therefore, we also developed a move visualisation process which attempts to identify all of the different types of moves that each player/piece can perform, and creates suitable images to demonstrate them.

4.1 Move Properties

For the purposes of move visualisation, there are four move properties that need to be considered:

1. **Mover:** Each move contains a *Mover* parameter, which indicates the player who is making the move. This property is only considered for games where the players have different piece rules.
2. **Piece:** The majority of moves contain an associated *Piece* parameter, which indicates the main piece that is moved, added, or removed. Even though a move may affect multiple pieces, a single piece is always designated as the main one. For example, when capturing a piece in chess there are technically two pieces that are being affected (the capturing piece and the captured piece), but the capturing piece is considered the main piece. Certain moves such as *Pass* or *Swap* do not have an associated piece.
3. **Origin Rules:** Each move originates from an associated (move ...) ludeme within the current game's description. For example, when playing the game description given in Fig. 1, all moves originate from (move Add (to (sites Empty))). By calling our previously described toEnglish() function on any move's associated ludeme, we can obtain an English translation of the rules from which this move originated.

4. **Action Types:** Each move contains a sequence of actions that are applied when the move is selected. For example, the move [(Remove E6), (Move F5-E6), (Score P1=4)] contains three actions which removes the piece at E6, moves the piece at F5 to E6, and sets the score of Player 1 to 4, respectively. The action types of a move are this same set of actions but where only the name/type of action is retained; for example the action types of the same example move given above would be [Remove, Move, Score].

4.2   Identifying Distinct Moves

When performing move visualisation for a given game, we first run a number of random playouts. Once completed, we then combine all of the moves that were selected during these playouts into a single list of moves. We then remove all duplicate moves from this list, using only the four move properties listed in Sect. 4.1 to determine uniqueness. This provides us with a set of distinct moves, each of which has a unique combination of properties.

4.3   Visual Representations

For each distinct move, we create two images showing the state of the game before and after the move is selected. The move's piece is also highlighted using either a red arrow or dot, depending on whether the piece's location changes or not. An example of these two images for a given move is shown in Fig. 3. Additionally, certain games within Ludii support move animations which can depict a moving piece in a more visually pleasing manner. In games where this animation is supported, we also provide a short GIF animation of the move.

5   Additional Features

As well as the English translation and move visualisation processes, there are several auxiliary features which can provide additional information.

5.1   Initial Setup

An image of the game state before any moves are made is also included. This is not very helpful for games where the board is empty, but can be beneficial for games such as Chess where the initial arrangement of the pieces is important.

5.2   Winning/Losing Moves

When running random playouts for move visualisation, we can also record the result of each playout as well as the final move that was made. We can then

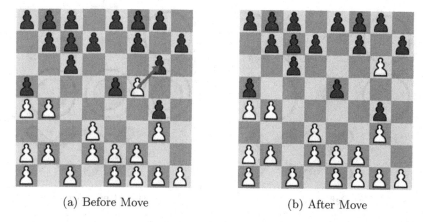

(a) Before Move   (b) After Move

Fig. 3. Visual representation of a move for Breakthrough, showing the board state before (a) and after (b) the move is selected. The move is highlighted using a red arrow. (Color figure online)

visualise this move for each unique result, showing the different possible results that the game can have as well as the last move that led to this outcome.

In addition to the regular move visualisation, we can also detect the specific (**end** ...) ludeme that caused the game to end. By calling our previously described toEnglish() function on this ludeme, we can obtain an English translation of the rules that ended the game. Certain (**end** ...) ludemes also support additional visuals, such as allowing us to identify the specific winning line or pattern. An example of these two ending move images for a result in Tic-Tac-Toe is shown in Fig. 4.

5.3 Similar Legal Moves

When highlighting moves using red arrows or dots, we can also highlight any other legal moves that have the same properties. Whereas previously only the selected move was shown, this approach displays more example moves within a single picture. Contrasting images showing with and without this addition are shown in Fig. 5. Both approaches have their benefits, however we feel that showing all similar legal moves more closely aligns with how piece rules are typically described in most Ludii game descriptions.

5.4 Strategy Explanation

Many of the games within Ludii also contain associated heuristics for assisting AI agents. The heuristics include aspects such as Material (number of pieces), Mobility (number of moves), LineCompletion (potential ability to complete lines of pieces), among many others. Each heuristic also has an associated weight which indicates its relative level of importance. These heuristics can be specified

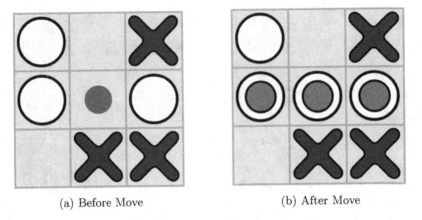

(a) Before Move   (b) After Move

Fig. 4. Visual representation of an ending move for Tic-Tac-Toe, showing the board state before (a) and after (b) the move is selected. The move is highlighted using a red dot, and the winning line is highlighted using green dots. (Color figure online)

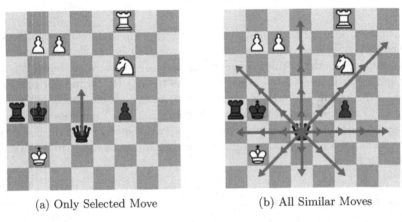

(a) Only Selected Move   (b) All Similar Moves

Fig. 5. Visual representation of the board state for Chess before a move is made, with only the selected move highlighted (a) and all legal moves with the same properties highlighted (b). (Color figure online)

manually by the game's designer, or learned automatically through a heuristic tuning process. More details on the heuristics which are available within Ludii can be found in [18]. While predominately designed for AI agents, these heuristics can also provide assistance to novice players about what moves to make. By applying some simple formatting, these heuristics can be converted into basic strategy explanations. An example of this strategy explanation for the Chess heuristics in Ludii is shown in Fig. 6.

```
Try to maximise the number of Pawn(s) you control (very low importance)
Try to maximise the number of Rook(s) you control (moderate importance)
Try to maximise the number of Bishop(s) you control (low importance)
Try to maximise the number of Knight(s) you control (low importance)
Try to maximise the number of Queen(s) you control (very high importance)
```

Fig. 6. Strategy explanation derived from Ludii's AI heuristics for Chess.

6   Complete Game Manuals

All of the aspects described in Sects. 3, 4, and 5 can be combined into a single webpage document which provides complete manuals for a specific game. Examples of generated manuals for a wide range of Ludii games are available online.[3] The layout of each manual is as follows:

1. **Rules:** The rules of the game, based on the English translation of its Ludii game description.
2. **Heuristics:** Recommended strategies for playing the game, derived from the AI heuristics.
3. **Setup:** The initial state of the game before any moves are made.
4. **Endings:** The different endings for the game, with accompanying English descriptions and visualisations.
5. **Moves:** The different moves for the game, with accompanying English descriptions and visualisations. These moves are hierarchically organised based on their properties (mover, piece, origin rules, and action types).

7   Conclusion

In this paper we have presented a process for automatically generating manuals of Ludii games. This process combines solutions for multiple sub-tasks to create a complete manual, detailing the rules, moves, endings and strategies for any given game. This manual allows players who are unfamiliar with L-GDL to still play complex games in Ludii, even if the hand-written rules are unavailable or incomplete. Procedurally generated games would be an ideal application of this work, as these games do not provide any instructions or details beyond their Ludii game description. These manuals may also be able to assist game designers, by providing a starting point for describing the game's rules or highlighting outlier moves that the designer may not have intended.

7.1   Limitations

While the manual generation process described in this paper works on the vast majority of games in Ludii, there are still some minor cases where specific parts

---

[3] https://ludii.games/manuals/menu.html.

are incompatible. Move visualisation is currently not possible for games with simultaneous moves, or games which reference smaller sub-games (aka. matches). The English translation code is also still in-progress, and will need to be updated and improved over time. Other minor aspects such as move animations and winning move visuals are also not yet implemented for all cases.

7.2 Future Work

One improvement on this work could be the creation of tutorial scenarios where a specific mechanic or rule needs to be understood in order to win, thus providing a more interactive learning experience [8]. Interactive tutorials have been shown to increase player engagement and ability, particularly for complex games [1,12]. These scenarios could be created by identifying game states where a certain move needs to be selected in order to win, or avoid loss.

Another improvement could be the addition of a coach/tutor AI that provides tips to the player throughout the game. This could include advice about what moves to make, or feedback on why certain moves were good/bad [11]. This could be achieved using our learned heuristics, by detecting if a different move would have given a higher state evaluation. For example, "Your last move resulted in a lower material benefit (+2) than an alternative move (+5)". Furthermore, features of state-action pairs [4] may be used to provide advice on a tactical level, as opposed to the strategic level that heuristic state functions operate on.

Our final improvement suggestion is the creation of an adaptive AI opponent which modifies its strength based on the player's abilities [10]. This agent would aim to provide a reasonable level of challenge for the player, hopefully providing a more engaging and constructive gameplay experience. This agent could be extended further to favour states which promote certain strategies, providing an online version of the tutorial scenarios mentioned previously.

Acknowledgements. This research is funded by the European Research Council as part of the Digital Ludeme Project (ERC Consolidator Grant #771292) led by Cameron Browne at Maastricht University's Department of Data Science and Knowledge Engineering.

Appendix

A   English Translations of Ludii Game Descriptions

A.1   Hex

```
(game "Hex"
    (players 2)
    (equipment{
        (board (hex Diamond 11))
        (piece "Marker" Each)
        (regions P1 { (sites Side NE) (sites Side SW) })
        (regions P2 { (sites Side NW) (sites Side SE) })})
    (rules
```

```
          (meta (swap))
          (play (move Add (to (sites Empty))))
          (end (if (is Connected Mover) (result Mover Win))))
)
```

The game "Hex" is played by two players on a 11x11 diamond board with hexagonal tiling.
Regions:
    RegionP1: the NE side for P1 and RegionP1: the SW side for P1
    RegionP2: the NW side for P2 and RegionP2: the SE side for P2
All players play with Markers.
Players take turns moving.
Rules:
    Add one of your pieces to the set of empty cells.
Aim:
    If the region(s) of the moving player are connected, the moving player wins.

A.2   Amazons

```
(game "Amazons"
    (players 2)
    (equipment{
            (board (square 10))
            (piece "Queen" Each (move Slide (then (moveAgain))))
            (piece "Dot" Neutral)})
    (rules
        (start{
                (place "Queen1" {"A4" "D1" "G1" "J4"})
                (place "Queen2" {"A7" "D10" "G10" "J7"})})
        (play
            (if (is Even (count Moves))
            (forEach Piece)
            (move Shoot (piece "Dot0"))))
        (end (if (no Moves Next) (result Mover Win))))
)
```

The game "Amazons" is played by two players on a 10x10 rectangle board with square tiling.
All players play with Queens. The following pieces are neutral: Dots.
Rules for Pieces:
    Queens slide from the location of the piece in the adjacent direction through the set of
        empty cells then move again.
Players take turns moving.
Setup:
    Place a Queen for player one on sites: A4, D1, G1 and J4.
    Place a Queen for player two on sites: A7, D10, G10 and J7.
Rules:
    If the number of moves is even, move one of your pieces, else shoot the piece Dot0.
Aim:
    If the next player cannot move, the moving player wins.

References

1. Andersen, E., et al.: The impact of tutorials on games of varying complexity, pp. 59–68. Association for Computing Machinery, New York (2012)
2. Björnsson, Y.: Learning rules of simplified boardgames by observing. In: Proceedings of the Twentieth European Conference on Artificial Intelligence, pp. 175–180 (2012)
3. Browne, C.: Modern techniques for ancient games. In: IEEE Conference on Computational Intelligence and Games, Maastricht, pp. 490–497. IEEE Press (2018)
4. Browne, C., Soemers, D.J.N.J., Piette, E.: Strategic features for general games. In: Proceedings of the 2nd Workshop on Knowledge Extraction from Games (KEG), pp. 70–75 (2019)

5. Browne, C.: A class grammar for general games. In: Plaat, A., Kosters, W., van den Herik, J. (eds.) CG 2016. LNCS, vol. 10068, pp. 167–182. Springer, Cham (2016). https://doi.org/10.1007/978-3-319-50935-8_16

6. Cropper, A., Evans, R., Law, M.: Inductive general game playing. Mach. Learn. **109**(7), 1393–1434 (2019). https://doi.org/10.1007/s10994-019-05843-w

7. Green, M.C., Khalifa, A., Barros, G.A.B., Machado, T., Nealen, A., Togelius, J.: AtDELFI: automatically designing legible, full instructions for games. In: Proceedings of the 13th International Conference on the Foundations of Digital Games, FDG 2018. Association for Computing Machinery, New York (2018)

8. Green, M.C., Khalifa, A., Barros, G.A.B., Nealen, A., Togelius, J.: Generating levels that teach mechanics. In: Proceedings of the 13th International Conference on the Foundations of Digital Games, FDG 2018. Association for Computing Machinery, New York (2018)

9. Gregory, P., Schumann, H.C., Björnsson, Y., Schiffel, S.: The *GRL* system: learning board game rules with piece-move interactions. In: Cazenave, T., Winands, M.H.M., Edelkamp, S., Schiffel, S., Thielscher, M., Togelius, J. (eds.) CGW/GIGA -2015. CCIS, vol. 614, pp. 130–148. Springer, Cham (2016). https://doi.org/10.1007/978-3-319-39402-2_10

10. Iida, H., Matsubara, H., Uiterwijk, J.: A search strategy for tutoring in game playing. In: IJCAI-1995 Workshop Proceedings, Entertainment and AI/Alife, pp. 14–18 (1995)

11. Ikeda, K., Viennot, S., Sato, N.: Detection and labeling of bad moves for coaching go. In: 2016 IEEE Conference on Computational Intelligence and Games (CIG), pp. 395–401 (2016)

12. Kelleher, C., Pausch, R.: Stencils-based tutorials: design and evaluation, pp. 541–550. Association for Computing Machinery, New York (2005)

13. Kowalski, J., Kisielewicz, A.: Regular language inference for learning rules of simplified boardgames. In: 2018 IEEE Conference on Computational Intelligence and Games (CIG), pp. 78–85 (2018)

14. Perez-Liebana, D., Liu, J., Khalifa, A., Gaina, R.D., Togelius, J., Lucas, S.M.: General video game AI: a multitrack framework for evaluating agents, games, and content generation algorithms. IEEE Trans. Games **11**(3), 195–214 (2019)

15. Perez-Liebana, D., Samothrakis, S., Togelius, J., Lucas, S.M., Schaul, T.: General video game AI: competition, challenges, and opportunities. In: Proceedings of the 30th AAAI Conference on Artificial Intelligence, pp. 4335–4337 (2016)

16. Piette, É., Browne, C., Soemers, D.J.N.J.: Ludii game logic guide (2021). https://arxiv.org/abs/2101.02120

17. Piette, É., Soemers, D.J.N.J., Stephenson, M., Sironi, C.F., Winands, M.H.M., Browne, C.: Ludii - the ludemic general game system. In: Giacomo, G.D., et al. (eds.) Proceedings of the 24th European Conference on Artificial Intelligence (ECAI 2020). Frontiers in Artificial Intelligence and Applications, vol. 325, pp. 411–418. IOS Press (2020)

18. Stephenson, M., Soemers, D.J.N.J., Piette, É., Browne, C.: General game heuristic prediction based on ludeme descriptions (2021). https://arxiv.org/abs/2105.12846

Optimised Playout Implementations for the Ludii General Game System

Dennis J. N. J. Soemers[(⊠)], Éric Piette, Matthew Stephenson, and Cameron Browne

Department of Data Science and Knowledge Engineering, Maastricht University, Paul-Henri Spaaklaan 1, 6229 EN Maastricht, The Netherlands
{dennis.soemers,eric.piette,matthew.stephenson, cameron.browne}@maastrichtuniversity.nl

Abstract. This paper describes three different optimised implementations of playouts, as commonly used by game-playing algorithms such as Monte-Carlo Tree Search. Each of the optimised implementations is applicable only to specific sets of games, based on their rules. The Ludii general game system can automatically infer, based on a game's description in its general game description language, whether any optimised implementations are applicable. An empirical evaluation demonstrates major speedups over a standard implementation, with a median result of running playouts 5.08 times as fast, over 145 different games in Ludii for which one of the optimised implementations is applicable.

Keywords: Playouts · General Game Playing · Ludii

1 Introduction

The playing strength of automated game-playing agents based on tree search algorithms, such as $\alpha\beta$-pruning [10] and Monte-Carlo Tree Search (MCTS) [3,8,11], typically correlates strongly with the efficiency of basic operations such as computing a list of legal moves, applying a move to a state, copying a game state, or evaluating whether or not a state is terminal. When such operations can be implemented to run more efficiently, they allow for deeper tree searches, which usually leads to stronger agents. For this reason, a significant amount of research has gone towards techniques such as bitboard methods [2], Prop-Net optimisations [17] for general game playing, hardware accelerators [6,18], optimising compilers for general game description languages [12], etc.

MCTS is one of the most commonly used tree search algorithms for general game playing [9,20]. Typically, a significant portion of the time spent by this algorithm is in running *playouts*; these may intuitively be understood as the algorithm following a "narrow" and "deep" trajectory of several—often many—consecutive states and actions. In their most basic form, playouts are run by selecting legal actions uniformly at random, and continuing them until a terminal game state is reached, but it is also possible to truncate playouts early and to select actions during playouts according to non-uniform distributions.

© Springer Nature Switzerland AG 2022
C. Browne et al. (Eds.): ACG 2021, LNCS 13262, pp. 223–234, 2022.
https://doi.org/10.1007/978-3-031-11488-5_20

After running a playout, it is typically not necessary to retain the intermediate states generated between the start and end of a playout, the lists of legal moves, etc.; only the final outcome of a playout is generally of interest. This is in contrast to minimax-based algorithms such as $\alpha\beta$-pruning [10], or even the time spent by MCTS in its tree building and traversal (outside of playouts), where intermediate states and exact lists of legal moves are required for a correct tree to be built. Straightforward playout implementations compute exact lists of legal moves in every state anyway, such that actions may be sampled from them afterwards, but these insights may be used to develop more efficient playout implementations.

In this paper, we propose several different optimised playout implementations for the Ludii general game system [4,14], which allow for playouts to be run significantly more quickly than with naive implementations. Each of them is only applicable to a restricted set of games, but the system can automatically determine for any given game whether or not any specific playout implementation is applicable. Furthermore, each of the proposed implementations is applicable to a substantial number of games in Ludii (i.e., not specific to just a single or a handful of games). Only our approach for automatically determining the applicability of playout implementations is specific to Ludii—in particular, to its game description format. The basic ideas behind the optimised playout implementations are not specific to Ludii, and may be relevant for other general game systems as well as single-game engines.

2 Background

Ludii is a general game system that can run any game described in its *ludemic* game description format [4,14]. A large library of *ludemes*, which may intuitively be understood as keywords that make up the game description language, is automatically inferred from Ludii's codebase using a class grammar approach [5]. An example game description for the game of Tic-Tac-Toe in Ludii's game description language is provided by Fig. 1.

Any game described in this language can be compiled by Ludii, resulting in a forward model with functions for computing lists of legal moves, applying moves to game states, copying game states, etc. Given these functions, a straightforward playout implementation can be written as in Algorithm 1.

Algorithm 1. Standard playout implementation.

Require: Game state s to start playout from.
1: **while** playout should be continued **do**   // Not terminal and not truncated
2:     legal_moves ← COMPUTELEGALMOVES(s)
3:     Sample move m from legal_moves   // Often uniformly at random
4:     Apply move m to state s
5: **end while**
6: **return** game state s at end of playout.

```
(game "Tic-Tac-Toe"
    (players 2)
    (equipment {
        (board (square 3))
        (piece "Disc" P1)
        (piece "Cross" P2)
    })
    (rules
        (play (move Add (to (sites Empty))))
        (end (if (is Line 3) (result Mover Win)))
    )
)
```

Fig. 1. Game description for Tic-Tac-Toe in Ludii's game description language.

3 Related Work

For several connection games [15] (and possibly other types of games), it can be proven that a game always ends in a win for exactly one player (no ties), and that the outcome does not change if play "continues" after reaching a terminal game state until the game board is full. For such games, playouts can be optimised by simply continuing them until the board is full, and only evaluating the outcome once at the end [15]. This is efficient because evaluating the win condition, which is often the most expensive computation of these games, only needs to be done once, at the end of every playout. This is in contrast to standard playout implementations as in Algorithm 1, where the win condition would be evaluated after every move.

In a general game system such as Ludii, we do not have a straightforward way to automatically prove or disprove for any arbitrary game description that the properties required for the optimisation described above hold. However, the techniques we propose in the following sections are similar in the sense that they are tailored specifically towards optimising playouts, as opposed to more generally optimising functions that are also used outside of playouts.

4 Add-to-Empty Playouts

The first collection of games for which we propose an optimised playout implementation is the set of games where players' moves consist of placing pieces of their colour on empty positions on a game board, and pieces can never be moved or removed anymore after being placed. We refer to these as "add-to-empty" games. This includes many well-known games such as *Gomoku, Havannah, Hex, Tic-Tac-Toe, Yavalath,* etc. These are often connection or line-completion games.

More formally, in Ludii, these games are recognised as those games where the playing rules are defined as (play (move Add (to (sites Empty)))). This is a strong restriction because only a single specific set of playing rules is permitted, but in practice we find this particular ruleset to be relatively commonly used among several popular games. For this specific set of rules, we are guaranteed that the list of legal moves in the initial game state is simply represented by all positions that are empty at the start of the game (generally the entire board), and that this list of legal moves monotonically decreases by exactly one after every move. This allows for an optimised implementation, where the list of legal moves is pre-allocated once at the start of a playout, and legal moves do not need to be re-computed at any later stage in the same playout.

The only exception that we implement additional support for is the *swap rule* (or pie rule). This is a common rule used in many of the games we aim to cover with this playout implementation, such as Hex and Havannah, which states that in the first turn of the second player, that player may opt to swap colours with their opponent, rather than making a move. This rule is intended to eliminate a first-mover advantage that the first player otherwise often has in these games. The presence of this rule technically means that the list of moves does not monotonically decrease by one in the very first turn transition, but it is straightforward to implement support for this one special case in the optimised playout implementation.

Note that, in these games, the idea of pre-computing a list of legal moves only once at the start, and monotonically removing moves as they are played afterwards, does not necessarily have to be restricted to just playouts. If such a list of moves were stored in memory in the game state representation, and updated as moves were applied, the optimisation could also be used outside of playouts (e.g., when building search trees). In the Regular Boardgames system [13], such an idea has been implemented more generally as a step of an optimising compiler [12]. However, we remark that this does increase the memory footprint of the game state representation, and it can slow down operations such as the copying of game states, which is often required in aspects of game tree searches outside of playouts.[1] By restricting the use of this idea to just playouts, where generating intermediate copies of game states is not required, we are guaranteed that it cannot inadvertently cause a slowdown.

5   Filter Playouts

The second collection of games for which we provide an optimised playout implementation is the set of games where there is a basic set of arbitrary rules that defines an initial list of legal moves for any game state s, but some of these moves m are afterwards filtered out if a certain postcondition fails for whichever successor state s' is reached if m were to be applied to s. A well-known example of such a game is *Chess*, where at first the moves are described according to the different move rules of different pieces, but any move m that would lead to a successor state s' where the mover's king would (still) be under threat is filtered out. In chess-specific engines, such conditions may be relatively cheap

---

[1] Ludii often requires game states to be copied during tree searches because Ludii does not support "undoing" moves, though this may be added in the future.

to compute without actually generating all the hypothetical successor states s'. However, in the Ludii general game system, these conditions are expensive to compute because all the potential successor states s' are fully generated (which in turn first requires many copies of s to be generated) to evaluate the postconditions.

More formally, we provide support for any game in Ludii where the playing rules are described in any one of the following formats, where isolated capital letters A, B, etc. can be filled by any arbitrary rules as permitted by the game description language:

1. `(play (do A ifAfterwards:(B)))`
2. `(play (if A B (do C ifAfterwards:(D))))`
3. `(play (or (do A ifAfterwards:(B)) (move Pass)))`

The first case is the most basic case, where A defines the rules used to generate the unfiltered list of moves, and B defines the postcondition that must hold in the successor state for any move generated by A not to be filtered out. The second case generates moves according to B if condition A holds, and otherwise drops into a similar construction as in the first case. This construction is frequently used in games such as *Chess* and *Shogi*, where promotion moves are generated if the player to make a move is the same player as the last mover, and regular moves with postconditions are generated otherwise. The third case is similar to the first case, except it also always generates an unconditional pass move as a legal move. This is used for games such as *Go*, where placing stones is conditional on liberty postconditions, but passing is always permitted. Other (more complex) cases than these three may occur and could be supported, but adding such support would require a small amount of additional engineering effort on a case-by-case basis. In practice we found these three cases to provide sufficient coverage for a substantial number of games, including several popular ones such as Chess, Go, and Shogi.

When constructing game trees, we cannot avoid computing the expensive postconditions, because the exact lists of legal moves must be fully generated to construct a correct game tree. However, in playouts, we only require the ability to sample legal moves according to some desired distribution over the legal moves, but do not necessarily need to know which other (unsampled) moves were actually legal according to the postconditions. Hence, we propose a playout implementation where moves are generated without checking postconditions. A rejection sampling approach is used where postconditions are evaluated only after a move has been selected (uniformly at random, in the simplest case), and the process is repeated if it turns out that the sampled move should have been filtered out. This allows us to avoid evaluating potentially expensive postconditions for moves that are not sampled. Pseudocode for this approach is provided by Algorithm 2. Section 7 discusses how this approach can be combined with more sophisticated playouts with non-uniform distributions over moves.

Algorithm 2. Optimised filter playout.

Require: Game state s to start playout from.
```
 1: while playout should be continued do        // Not terminal and not truncated
 2:     moves ← ComputeMaybeLegalMoves(s)        // Ignore postconditions
 3:     m ← Null
 4:     while m = Null do
 5:         m ← sample move m from moves
 6:         if m fails postcondition then
 7:             m ← Null
 8:             Remove m from moves
 9:         end if
10:     end while
11:     Apply move m to state s
12: end while
13: return game state s at end of playout.
```
---

6   No-Repetition Playouts

The final playout implementation we propose is a variant of the filter playouts described in the previous section. Outside of the general playing rules, Ludii's game description language also allows for a more general (noRepeat) "meta-rule" to be applied to a complete game. When this rule is used, any move that leads to a game state that has already been encountered before is illegal. This can be viewed as an additional postcondition, which again requires a game state copy and a move application to evaluate, as described in Sect. 5. A similar rejection sampling approach can also be used again to avoid these computations for many legal moves in playouts. The main difference between the no-repetition playout and the filter playout is simply in how its applicability can be determined from a game's game description file. In games where filter playouts are also valid, any repetition restrictions are evaluated at the same time as the optimised postconditions.

7   Non-uniform Move Distributions

Selecting moves uniformly at random is a common and straightforward strategy, but it is often beneficial to use "smarter" playouts based on domain knowledge, offline learning, or online learning, which means that moves are sampled according to non-uniform distributions over the legal moves. The add-to-empty playouts described in Sect. 4 still generate the precise lists of legal moves, which means that they support the use of such non-uniform distributions. However, the filter playouts and no-repetition playouts described in Sects. 5 and 6 require careful attention. These playout implementations may include illegal moves in their lists of moves, which are only discovered to be illegal and rejected after sampling them, but their presence in the initial list of moves may affect the probabilities computed for other (legal) moves. This may lead to an unintended change in the distribution over moves.

One common approach for move selection in playouts is to assign scores to moves, which are not translated into probabilities, but instead used to inform move selection through other means, such as ϵ-greedy policies. An ϵ-greedy strategy simply selects moves uniformly at random with probability $0 \leq \epsilon \leq 1$, or greedily with respect to the move scores with probability $1 - \epsilon$. Move scores can, for example, be obtained using approaches such as MAST, FAST [9], NST [21], or PPA [7]. Techniques with only two or three discrete levels of prioritisation for moves, such as the Last-Good-Reply policy [1] or decisive and anti-decisive moves [22], may be viewed as a special case with discrete move scores. Whenever such an ϵ-greedy policy is used (including the special case of greedy policies with $\epsilon = 0$), our proposed playout implementations—with their rejection sampling schemes for handling illegal moves—will automatically play according to the correct (non-uniform) distributions, with no further changes required.

Another common approach is to compute a discrete probability distribution over all moves, and sample moves according to those probabilities. This is sometimes done by transforming move scores, such as those described above, into probabilities using a Boltzmann distribution. Given a set of legal moves $\mathcal{M}$, and a temperature hyperparameter τ, the probability $p(m, \mathcal{M})$ with which a move $m \in \mathcal{M}$ with a score $Q(m)$ should be selected is then given by Eq. 1:

$$p(m, \mathcal{M}) = \frac{\exp(Q(m)/\tau)}{\sum_{m' \in \mathcal{M}} \exp(Q(m')/\tau)} \tag{1}$$

When offline training is used to train policies, for instance based on deep neural networks [16] or simpler function approximators and state-action features [19], it is also customary to use such a distribution with $\tau = 1$ (leading to a softmax distribution) and the $Q(m)$ values referred to as logits.

Let $\mathcal{M}$ denote a set of legal moves, and let $\mathcal{I}$ denote a set of moves as generated during a filter or no-repetition playout (which may include some illegal moves), such that $\mathcal{M} \subseteq \mathcal{I}$. Let m_1 and m_2 denote two arbitrary legal moves. The ratio $\frac{p(m_1, \mathcal{I})}{p(m_2, \mathcal{I})}$ between their probabilities, in the possible presence of illegal moves, is given by Eq. 2:

$$\frac{p(m_1, \mathcal{I})}{p(m_2, \mathcal{I})} = \frac{\exp(Q(m_1)/\tau)}{\sum_{m' \in \mathcal{I}} \exp(Q(m')/\tau)} \times \frac{\sum_{m' \in \mathcal{I}} \exp(Q(m')/\tau)}{\exp(Q(m_2)/\tau)} = \frac{\exp(Q(m_1)/\tau)}{\exp(Q(m_2)/\tau)} \tag{2}$$

Note that this ratio is equal to the ratio we would have had with $\mathcal{M}$ instead of $\mathcal{I}$, i.e. if there were no possible presence of illegal moves.

Let $m \in \mathcal{I}$ denote a move that has been sampled in a playout, and is rejected due to it turning out to be illegal, i.e. $m \notin \mathcal{M}$. For any other move $m' \neq m$, the probability value $p(m', \mathcal{I} \setminus \{m\})$ can be incrementally updated as $p(m', \mathcal{I} \setminus \{m\}) = p(m', \mathcal{I}) \times \frac{1}{1 - p(m, \mathcal{I})}$ when m is rejected. This re-normalises the distribution into a proper probability distribution again after the rejection of the illegal move, without changing the ratio of probabilities between any pair of remaining moves, and without requiring the full distribution to be re-computed from scratch.

Table 1. Aggregate measures of the speedups obtained by different playout implementations in their applicable games.

Playout Implementation	Num. Games	Speedup			
		Min	Median	Mean	Max
Add-To-Empty	35	1.00	1.90	3.64	20.25
Filter	105	1.18	5.49	6.88	34.31
No-Repetition	5	1.65	6.35	9.08	19.26
All	145	1.00	5.08	6.17	34.31

8 Empirical Evaluation

We evaluate the performance of the proposed playout implementations by measuring the average number of complete random playouts—from initial game state until terminal game state—that can be run per second, using both standard implementations (Algorithm 1) and the optimised implementations. Every process is run on a single CPU core @2.2 GHz, using 60 s of warming up time for the Java Virtual Machine (JVM), followed by 600 s over which the number of playouts run per second is measured. We allocate 5120 MB of memory per process, of which 4096 MB is made available to the JVM.

The version of Ludii used for this evaluation[2] has 929 different games, with 1053 rulesets (some games can be played using several different variants of rules). Of these, 145 rulesets (from 141 games) are automatically detected to be compatible with one of the three proposed playout implementations. For each of them, we evaluate the speedup as the number of playouts per second when using the optimised playout, divided by the number of playouts per second when using a standard playout implementation. For example, a speedup of 2.0 means that the optimised implementation allows for playouts to be run twice as fast.

Figure 2 summarises, for each of the three playout implementations, the different speedups obtained by using the optimised playout implementations in applicable games. Table 1 provides additional details on these results. Each of the three implementations provides noticeable speedups in the majority of games, with median speedups ranging from 1.90 (almost twice as fast) for Add-To-Empty, to 6.35 (more than six times faster) for No-Repetition. The largest speedup (34.31) is obtained by the Filter playout in the game of *Go*.

Only the Add-To-Empty playout has two games (out of 35) for which the speedup is lower than 1.0, i.e. a slowdown; 0.9999896 for *Icosian*, and 0.997 for *Gyre*. In Icosian, the Add-To-Empty playout is only valid for the first phase of the game, which only lasts for a single move; after this phase, it is necessary to switch back to the standard playout implementation, and the overhead of this switch may cause the slowdown. In Gyre, close to 100% of the time is spent computing the game's win condition, which is not affected by Add-To-Empty.

---

[2] Revision 7903697 of https://github.com/Ludeme/Ludii.

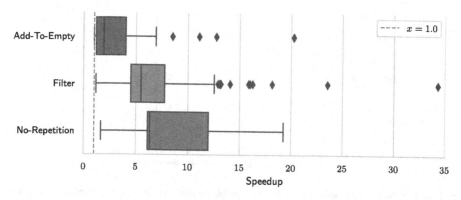

Fig. 2. Boxplots summarising speedups obtained from using optimised playout implementations rather than the standard one. Every data point is a different game (or ruleset). Points to the left of the $x = 1.0$ line are slowdowns.

In theory, the optimised playout implementations should not affect the probabilities with which moves are selected, and therefore random playouts should—on average—take equally long (measured in number of moves per playout) regardless of implementation. To verify that this is the case (i.e., there are no implementation errors), we compute a ratio for every game by dividing the average playout length recorded when using optimised implementations, by the corresponding number recorded when using the standard (unoptimised) implementation. The boxplots in Fig. 3 confirm that almost all these ratios are very close to 1.0.

The three biggest outliers are *Hexshogi*, *Unashogi*, and *Yonin Shogi*, with ratios of 0.75, 0.87, and 1.13, respectively. All three of these games are relatively slow games, which means that even in our 600-s timing runs we obtain relatively low total numbers of playouts, with a significant variance in the number of moves per playout. Therefore, the observation of these outliers can be explained by a combination of relatively low sample sizes (459, 215, and 322 total playout counts over 600 s for the three respective games when using optimised playout implementations) and high variance, rather than implementation errors. For all three of these outliers, the speedups recorded for the Filter playout are also more substantial than can be explained solely by the differences in average playout lengths; we record speedups of 7.52, 5.67, and 4.50.

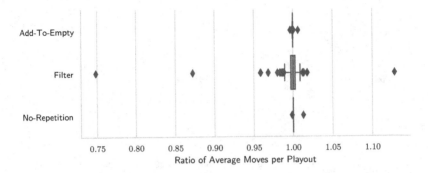

Fig. 3. For each of the optimised playout implementation, a boxplot summarising, for each game, the ratio between the recorded average numbers of moves per random playout with and without using the optimised implementation. Ratios less than 1.0 mean that random playouts were shorter on average when using optimised implementations, and ratios greater than 1.0 mean that random playouts were longer on average when using optimised implementations.

9   Conclusion

In this paper, we have proposed three optimised implementations for running playouts, as often used by algorithms such as MCTS. Each of the implementations is applicable to a specific set of games, depending on the rules used by a game. The Ludii general game system can automatically infer, based on game descriptions in its game description language, which—if any—of these implementations are applicable, and use them for running playouts when applicable. An empirical evaluation across 145 games demonstrated significant speedups, with a median result of running playouts 5.08 times faster, a mean speedup of 6.17 times, and a maximum speedup of 34.31 times in the game of Go.

Acknowledgements. This research is funded by the European Research Council as part of the Digital Ludeme Project (ERC Consolidator Grant #771292) led by Cameron Browne at Maastricht University's Department of Data Science and Knowledge Engineering. We thank the anonymous reviewers for their feedback.

References

1. Baier, H., Drake, P.D.: The power of forgetting: improving the last-good-reply policy in Monte Carlo go. IEEE Trans. Comput. Intell. AI Games **2**(4), 303–309 (2010)
2. Browne, C.: Bitboard methods for games. ICGA J. **37**(2), 67–84 (2014)
3. Browne, C., et al.: A survey of Monte Carlo tree search methods. IEEE Trans. Comput. Intell. AI Games **4**(1), 1–49 (2012)
4. Browne, C., Stephenson, M., Piette, É., Soemers, D.J.N.J.: A practical introduction to the Ludii general game system. In: Cazenave, T., van den Herik, J., Saffidine, A., Wu, I.C. (eds.) Advances in Computer Games. LNCS, vol. 12516, pp. 167–179. Springer, Cham (2020). https://doi.org/10.1007/978-3-030-65883-0_14

5. Browne, C.: A class grammar for general games. In: Plaat, A., Kosters, W., van den Herik, J. (eds.) CG 2016. LNCS, vol. 10068, pp. 167–182. Springer, Cham (2016). https://doi.org/10.1007/978-3-319-50935-8_16

6. Campbell, M., Joseph Hoane Jr., A., Hsu, F.: Deep blue. Artif. Intell. **134**(1–2), 57–83 (2002)

7. Cazenave, T.: Playout policy adaptation for games. In: Plaat, A., van den Herik, J., Kosters, W. (eds.) ACG 2015. LNCS, vol. 9525, pp. 20–28. Springer, Cham (2015). https://doi.org/10.1007/978-3-319-27992-3_3

8. Coulom, R.: Efficient selectivity and backup operators in Monte-Carlo tree search. In: van den Herik, H.J., Ciancarini, P., Donkers, H.H.L.M.J. (eds.) CG 2006. LNCS, vol. 4630, pp. 72–83. Springer, Heidelberg (2007). https://doi.org/10.1007/978-3-540-75538-8_7

9. Finnsson, H., Björnsson, Y.: Learning simulation control in general game-playing agents. In: Proceedings of 24th AAAI Conference on Artificial Intelligence, pp. 954–959. AAAI Press (2010)

10. Knuth, D.E., Moore, R.W.: An analysis of alpha-beta pruning. Artif. Intell. **6**(4), 293–326 (1975)

11. Kocsis, L., Szepesvári, C.: Bandit based Monte-Carlo planning. In: Fürnkranz, J., Scheffer, T., Spiliopoulou, M. (eds.) ECML 2006. LNCS (LNAI), vol. 4212, pp. 282–293. Springer, Heidelberg (2006). https://doi.org/10.1007/11871842_29

12. Kowalksi, J., et al.: Efficient reasoning in regular boardgames. In: Proceedings of 2020 IEEE Conference on Games, pp. 455–462. IEEE (2020)

13. Kowalski, J., Maksymilian, M., Sutowicz, J., Szykuła, M.: Regular boardgames. In: Proceedings of 33rd AAAI Conference on Artificial Intelligence, pp. 1699–1706. AAAI Press (2019)

14. Piette, É., Soemers, D.J.N.J., Stephenson, M., Sironi, C.F., Winands, M.H.M., Browne, C.: Ludii - the ludemic general game system. In: Giacomo, G.D., et al. (eds.) Proceedings of the 24th European Conference on Artificial Intelligence (ECAI 2020). Frontiers in Artificial Intelligence and Applications, vol. 325, pp. 411–418. IOS Press (2020)

15. Raiko, T., Peltonen, J.: Application of UCT search to the connection games of Hex, Y, *Star, and Renkula! In: Proceedings of Finnish Artificial Intelligence Conference, pp. 89–93 (2008)

16. Silver, D., et al.: A general reinforcement learning algorithm that masters chess, shogi, and Go through self-play. Science **362**(6419), 1140–1144 (2018)

17. Sironi, C.F., Winands, M.H.M.: Optimizing propositional networks. In: Cazenave, T., Winands, M.H.M., Edelkamp, S., Schiffel, S., Thielscher, M., Togelius, J. (eds.) CGW/GIGA -2016. CCIS, vol. 705, pp. 133–151. Springer, Cham (2017). https://doi.org/10.1007/978-3-319-57969-6_10

18. Siwek, C., Kowalski, J., Sironi, C.F., Winands, M.H.M.: Implementing propositional networks on FPGA. In: Mitrovic, T., Xue, B., Li, X. (eds.) AI 2018. LNCS (LNAI), vol. 11320, pp. 133–145. Springer, Cham (2018). https://doi.org/10.1007/978-3-030-03991-2_14

19. Soemers, D.J.N.J., Piette, É., Browne, C.: Biasing MCTS with features for general games. In: Proceedings of 2019 IEEE Congress on Evolutionary Computation, pp. 442–449. IEEE (2019)

20. Świechowski, M., Park, H., Mańdziuk, J., Kim, K.J.: Recent advances in general game playing. Sci. World J. (2015)

21. Tak, M.J.W., Winands, M.H.M., Björnsson, Y.: N-grams and the last-good-reply policy applied in general game playing. IEEE Trans. Comput. Intell. AI Games 4(2), 73–83 (2012)
22. Teytaud, F., Teytaud, O.: On the huge benefit of decisive moves in Monte-Carlo tree search algorithms. In: Proceedings of the IEEE Symposium on Computational Intelligence and Games, Dublin, pp. 359–364 (2010)

General Board Geometry

Cameron Browne[(✉)], Éric Piette, Matthew Stephenson,
and Dennis J. N. J. Soemers

Department of Data Science and Knowledge Engineering, Maastricht University,
Paul-Henri Spaaklaan 1, 6229 EN Maastricht, The Netherlands
{cameron.browne,eric.piette,matthew.stephenson,
dennis.soemers}@maastrichtuniversity.nl

Abstract. Game boards are described in the Ludii general game system by their underlying graphs, based on tiling, shape and graph operators, with the automatic detection of important properties such as topological relationships between graph elements, directions and radial step sequences. This approach allows most conceivable game boards to be described simply and succinctly.

Keywords: General Game Playing · Ludii · game board · geometry

1 Introduction

The Digital Ludeme Project (DLP) is a five-year research project using Artificial Intelligence techniques to improve our understanding of the development of games throughout history [2]. We are modelling the 1,000 most "important" traditional strategy games in a consistent digital format, to provide a playable database of the world's traditional games for comparative analysis.

The Ludii general game system[1] [5] is a software tool developed specifically for this task, for modelling the full range of possible board games (950+ games implemented in version 1.2.8). Games are described in terms of simple *ludemes* assembled into structures to define arbitrarily complex behaviour, where each ludeme is a game-related concept implemented as a Java class (or enum attribute) in the Ludii code base [4].

A key challenge in this task is to allow the user to describe arbitrarily complex game boards in a simple and intuitive way. This paper outlines our method for describing game boards in the Ludii grammar for general games.

2 Game Graphs

In Ludii, the board shared by all players is represented internally as a finite graph defined by a triple of sets $\mathcal{G} = \langle V, E, C \rangle$ in which V is a set of *vertices*, E a set of *edges*, and C a set of *cells*. In graph theory, a cell is more commonly

---

[1] Ludii is available at ludii.games and the source code at github.com/Ludeme/Ludii.

© Springer Nature Switzerland AG 2022
C. Browne et al. (Eds.): ACG 2021, LNCS 13262, pp. 235–246, 2022.
https://doi.org/10.1007/978-3-031-11488-5_21

called *face* and represents a region bounded by a set of edges and that contains no other vertex or edge.[2] Vertex, edge and cell are all graph elements which can refer to each other, and denote *playable sites* at which players can place components during the game:

- Let $v \in V$ denote a vertex. Then v is an endpoint to each edge in $E(v)$, $C(v)$ gives the set of cells that v is part of, and $V(v) = \{v\}$.
- Let $e \in E$ denote an edge. Then $V(e)$ is a set of 2 vertices that are the endpoints of e, $C(e)$ gives the set of cells e is bounding, and $E(e) = \{e\}$.
- Let $c \in C$ denote a cell. Then $E(c)$ is the set of all the edges bounding c, $V(c)$ gives the set of the vertices which are the endpoints of the edges bounding c, and $C(c) = \{c\}$.

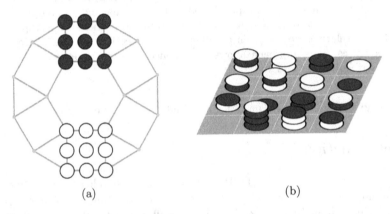

(a)   (b)

Fig. 1. A game played on vertices, edges and cells (a) and a game played on cells (b).

For example, Fig. 1a shows a game with pieces played on the vertices, edges and cells of the board graph. Figure 1b shows a board game played only on the cells but in which pieces may stack.

In any single game, components (or a stack of components) can be placed on any graph element. For this reason, we define a playable site as a triple $\langle Type, Index, Level \rangle$ in which the *Type* can be (Vertex, Edge or Cell), the *Index* is the number of the element and the *Level* is the index of the element in the stack (0 meaning the ground).

Any ludeme referring to a playable site has to specify each of these data. However, for convenience, Ludii uses default values. The default type of a location is Cell, except if the description of the game specifies another default site type. For levels, the default value is the top Level of the location specified, as stacked site are typically owned by the player with a piece on top.

---

[2] We sometimes use "game design" terms or definitions in lieu of stricter mathematical equivalents, in keeping with Ludii's primary purpose as a game design tool.

2.1 Dimensions: Cells or Vertices

The graph is generated based on the specified board dimensions and default
site type. For example, a Chess board described (board (square 8)) (see
Fig. 2a) produces a square grid with 8 cells per row and column. However, a
Go board described as (board (square 19) use: Vertex) (see Fig. 2b) produces a square grid with 19 vertices per row and column. If the default site type
is Vertex or Edge then the board dimensions are based on the number of vertices
rather than cells.

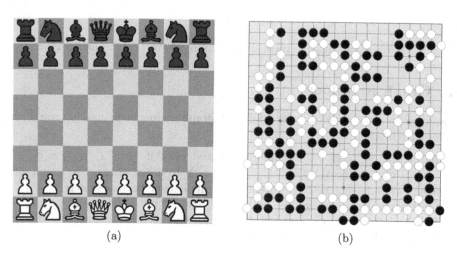

(a)   (b)

Fig. 2. (a) Chess (8 × 8 Cells). (b) Go (19 × 19 Vertices).

3 Game Board Description

Game boards are described in the Ludii grammar [3] using the following basic
EBNF syntax: <board> ::= (board <graph>) where the underlying <graph>
object defines the vertices, edges and cells that make up the game board.

The user can specify the location of each vertex (and adjacencies between
them as edges) to allow the description of arbitrarily complex graphs, or they
can take advantage of a range of predefined tilings, shapes and graph operators
for more concise descriptions (described more fully in Sect. 5). For example, the
three game boards shown in Fig. 3 are described by the following graphs (the
poly field describes the polygonal shape of the board):

```
(hex 4)
```

```
(tiling T3464 2)
```

```
(celtic (poly {{3 0}{3 4}{0 4}{0 7}{3 7}{3 11}{6 11}
          {6 7}{10 7} {10 5} {6 5}{6 0}}))
```

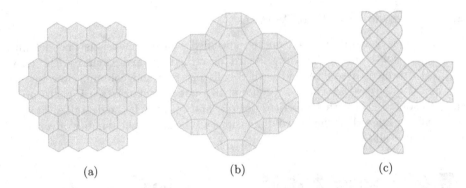

Fig. 3. Boards from tilings: hexagonal (a), semi-regular 3.4.6.4 (b) and `celtic` (c).

4   Graph Relations

For a graph $\mathcal{G} = \langle V, E, C \rangle$, two different graph elements g_1 and g_2 can have different relations:

- **Adjacent**: g_1 and g_2 are *adjacent* if and only if $(\exists e \in E(g_1) \cap E(g_2)) \vee (\exists v \in V(g_1) \cap V(g_2)) \vee (\exists c \in C(g_1) \cap C(g_2))$. In other words, two graph elements are adjacent if they share any graph element they are referring.
- **Orthogonal**: g_1 and g_2 are *orthogonal* if and only if $\exists e_1 \in E(g_1), \exists e_2 \in E(g_2), e_1 = e_2$. In other words, two graph elements are orthogonal if they share an edge.

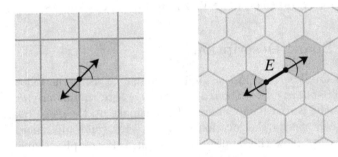

Fig. 4. Adjacent diagonals (left) and non-adjacent diagonals (right).

- **Diagonal**: Two cells are considered *diagonal* if and only if:[3]
  1. They share a vertex (but not an edge) and the bisectors of the angles at that vertex in each cell are maximally opposed. Note that a cell can have multiple diagonal neighbours through a vertex if all satisfy this property. Or:

---

[3] This definition differs slightly from the actual implementation, but it captures the general understanding of *diagonality* between cells on a game board.

2. If a cell has no such adjacent neighbour through a given vertex, then we allow a non-adjacent diagonal neighbour through that vertex if the two cells are coincident with the end points of some edge E (which does not belong to either cell) and the bisectors of the angles at the end point in each cell are maximally opposed.

These two diagonal relationships are shown in Fig. 4.

Diagonality is defined similarly for vertices, but transposing "cell" and "vertex" in the above definitions.

- **Off Diagonal**: g_1 and g_2 are off diagonal if and only if $g_1 \in C, g_2 \in C, \exists v_1 \in V(g_1), \exists v_2 \in V(g_2), v_1 = v_2, \nexists e_1 \in E(g_1), \nexists e_2 \in E(g_2), e_1 = e_2$. In other words, two cells are off diagonal if they are not diagonal, not orthogonal and they share a vertex.
- **All**: g_1 and g_2 are related if they are orthogonally, adjacently, diagonally or off diagonally related to each other.

These relationships are summarised for the regular tilings in Table 1.

4.1   Directions

Ludii supports the following direction types:

- *Intercardinal* directions: N, NNE, NE, ENE, E, ESE, SE, SSE, S, SSW, SW, WSW, W, WNW, NW, NNW.
- *Rotational* directions: In, Out, CW (clockwise), CCW (counter-clockwise).
- *Spatial* directions for 3D games: D, DN, DNE, DE, DSE, DS, DSW, DW, DNW and U, UN, UNE, UE, USE, US, USW, UW, UNW.
- *Axial* directions subset (for convenience): N, E, S, W.
- *Angled* directions subset (for convenience): NE, SE, SW, NW.

Each graph element g has a corresponding set of *absolute directions* A_d and *relative directions* R_d to associated graph elements of the same type. Absolute directions can be any of the above direction types in addition to any relation type (Adjacent, Orthogonal, Diagonal, Off Diagonal, or All).

Relative directions from an element g are defined by $R_d(g, facing, rotation, relation)$ where *facing* describes the direction in which a component at g is facing, *rotation* describes the number of rightward steps of the component at g, and *relation* describes the graph relation to use at each step (Adjacent by default). Relative directions are: Forward, Backward, Rightward, Leftward, FR, FRR, FRRR, FL, FLL, FLLL, BR, BRR, BRRR, BL, BLL or BLLL.

For example, consider a piece on a square board (which involves only the eight major compass directions as adjacent relations). If the piece is facing N (North) with a rotation of 0, the relative direction Forward is the graph element immediately to the North (upwards) if such an element exists. However, if that piece is facing E (East) and its current rotation is 1, the relative direction FR (meaning "Forward Right") is the graph element to its South East (if such an element exists).

Table 1. Relations for the regular tilings.

Relation	Square	Triangular	Hexagonal
All			
Adjacent			
Orthogonal			
Diagonal			
Off-Diagonal			

4.2 Steps and Walks

A *step* is a record of two related graph elements (*from* and *to*) which can be of different types and the absolute directions that describe their relationship. For example, a cell A directly above another cell B on a Chess board could be described as an `Adjacent`, `Orthogonal` or `N` step away.

Ludii also provides three relative step types (`F`, `L` and `R`) that allow users to define *walks* within the board graph. These correspond the standard "forward", "left" and "right" commands used in *turtle graphics* [1], as shown in Fig. 5.

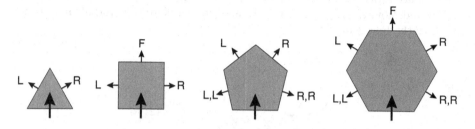

Fig. 5. Relative steps from various cell types.

This representation allows descriptions of piece movements to be easily transferred between different board topologies. For example, a knight move in Chess may be described as the walk {F,F,R} as shown in Fig. 6 (left) ans this walk may be directly used on a board based on the semi-regular 3.4.6.4 tiling (Fig. 6, right). Note, however, that different topologies may introduce ambiguities such as whether both right turns in the 3.4.6.4 knight move (dotted lines) should be considered valid moves or only one of them (probably the furthest reaching one). Such ambiguities should be resolved by the game designer according to the behaviour they want.

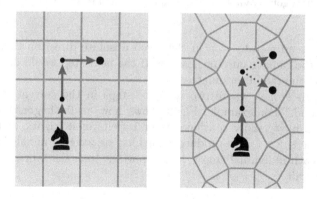

Fig. 6. Walk {F,F,R} describes knight moves on square and 3.4.6.4 tilings.

4.3 Radials

Many games involve piece movement through contiguous lines of cells in a direction, such as slide moves by the queen, rook and bishop pieces in Chess. Such lines of play are called *radials*. Ludii automatically pre-generates all possible steps between playable sites on the board and all possible radials derived from them, for convenient game description and efficient processing.

For each playable site on the board S, each valid step to a neighbouring graph element of the same type in an absolute direction d is extended as far as possible, to produce a radial from S in direction d. For example, Fig. 7 shows Orthogonal radials from the shaded cell on a circular Chess board, such as a rook would move in the game Shatranj ar-Rumiya. Note that radials may bend to follow the board topology.

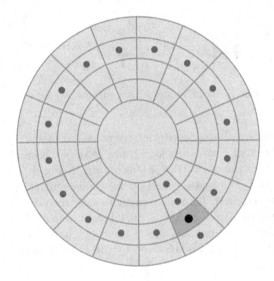

Fig. 7. Orthogonal radials on a circular Chess board (Shatranj ar-Rumiya).

Radials extend step-by-step in the given absolute direction that minimises deviation in the radial's current heading. If the next step would deviate by 90° or more, then the radial terminates.

Radials can *branch* where two or more steps in the current direction are equally as good.[4] For example, Fig. 8 shows how an Orthogonal step into a triangular cell may validly continue either L (left) or R (right), and thereafter alternate {L,R,L,R, . . . } to produce branching zig-zagging radials in which the direction of each individual step is less important than the average direction of the radial overall.

---

[4] Still to be implemented in Ludii.

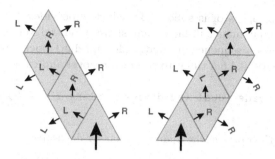

Fig. 8. Branching radial on a triangular grid.

5   Graph Operators

Graphs are initially defined by a tiling and/or shape but can then be further modified using a range of *graph operators*. The complete set of tilings, shapes and graph operators defined in the Ludii grammar is shown in Table 2. These can be used in combination to define thousands of different boards types quickly and easily. Greyed out items indicate planned future work not implemented yet.

For example, the very useful `dual` operator converts a source graph into its *weak dual* defined by edges whose end points are the centroids of its adjacent cells. Figure 9 shows a `dual` operation applied to a small graph based on tiling 3.3.4.3.4 to produce the well known Cairo tiling:

(dual (tiling T33434 2))

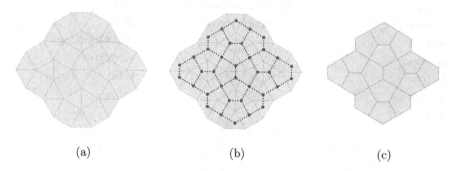

(a)   (b)   (c)

Fig. 9. A 3.3.4.3.4 tiling (a), its cell adjacencies (b) and its weak dual (c).

Another useful operator is subdivide, which subdivides all faces with N or more sides into triangular sub-faces that share a central vertex (default $N = 1$). Figure 10 shows a sequence of subdivide and dual operations applied to a rhombitrihexahedral 3.4.6.4 tiling to produce a novel and exotic board design:

```
(dual (subdivide (dual (subdivide (tiling T3464 2) min:6))))
```

Table 2. Keywords in the Ludii grammar for describing game boards.

Tiling	Shape	Operator
Regular	square	add
square	rectangle	clip
hex	hexagon	complete
tri	triangle	dual
	wedge	hole
Semi-Regular	regular (polygon)	intersect
T488 (*i.e.* 4.8.8)	poly (any polygon)	keep
T4612 (*i.e.* 4.6.12)		layers
T3464 (*i.e.* 3.4.6.4)	*Attribute*	makeFaces
T3636 (*i.e.* 3.6.3.6)	Star	merge
T31212 (*i.e.* 3.12.12)	Diamond	recoordinate
T33336 (*i.e.* 3.3.3.3.6)	Prism	remove
T33344 (*i.e.* 3.3.3.4.4)		renumber
T33434 (*i.e.* 3.3.4.3.4)	*Modifier*	rotate
	diagonals:<DiagType>	scale
Custom	pyramidal:<boolean>	shift
concentric	limping:<boolean>	skew
spiral	fractal/recursive	splitCrossings
quadhex	lattice	subdivide
brick	projective	trim
celtic		union
repeat		

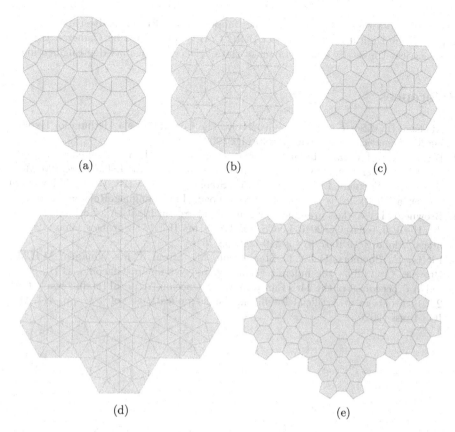

(a)   (b)   (c)

(d)   (e)

Fig. 10. A 3.4.6.4 tiling (a) subdivided at $N \geq 6$ (b), its dual (c), all subdivided (d) and its dual (e).

6 Conclusion

The Ludii grammar provides a simple way to describe most conceivable game boards by their underlying graphs, using tiling, shape and graph operators. This approach has allowed us to model the boards of over a thousand games for the Ludii general game system, and continues to produce interesting new board designs based on simple operations.

Future work will include improvements to symmetric board colourings (for games in which cell colour is relevant) and adaptive coordinate labelling that follows the contours of exotic boards. But the inclusion of a freeform **graph** ludeme means that almost any game board that can described as a combination of vertices, edges and/or cells can be defined, making this approach ideal for the wide range of game boards to be modelled for the Digital Ludeme Project.

Acknowledgements. This research was funded by the European Research Council as part of the Digital Ludeme Project (ERC CoG #771292) led by Cameron Browne at Maastricht University's Department of Data Science and Knowledge Engineering.

References

1. Abelson, H., diSessa, A.: Turtle Geometry: The Computer as a Medium for Exploring Mathematics. MIT Press, Cambridge (2008)
2. Browne, C.: Modern techniques for ancient games. In: IEEE Conference on Computational Intelligence and Games, Maastricht, pp. 490–497. IEEE Press (2018)
3. Browne, C., Soemers, D.J.N.J., Piette, É., Stephenson, M., Crist, W.: Ludii language reference (2020). https://ludii.games/downloads/LudiiLanguageReference.pdf
4. Browne, C.B.: A class grammar for general games. In: Plaat, A., Kosters, W., van den Herik, J. (eds.) CG 2016. LNCS, vol. 10068, pp. 167–182. Springer, Cham (2016). https://doi.org/10.1007/978-3-319-50935-8_16
5. Piette, É., Soemers, D.J.N.J., Stephenson, M., Sironi, C.F., Winands, M.H.M., Browne, C.: Ludii - the ludemic general game system. In: Giacomo, G.D., et al. (eds.) Proceedings of the 24th European Conference on Artificial Intelligence (ECAI 2020). Frontiers in Artificial Intelligence and Applications, vol. 325, pp. 411–418. IOS Press (2020)

Quantifying the Space of Hearts Variants

Mark Goadrich(✉) and Collin Shaddox

Hendrix College, Conway, AR 72032, USA
{goadrich,shaddoxac}@hendrix.edu

Abstract. Hearts is a card game with a rich history and many interesting variants. Why has it remained popular while undergoing significant changes? We use computational simulations of Hearts to understand the experience of players through the application of four heuristics which quantify the drama and security felt by the winning player, the ability of players to win in chaotic imperfect-information situations, and the player's ultimate interest in their decisions. We find that there is a direct relationship between the historical evolution of Hearts through ludemic change and subsequent heuristic improvements to game play.

Keywords: Card Games · General Game Playing · Heuristics

1 Introduction

A ludeme [15] "is a fundamental unit of play," also known as a building block or mechanic by which game rules are constructed [8]. Over time, in a similar manner to living organisms, games can evolve through changes to their underlying ludemic structure [5], through rearrangement, additions, deletions, and mutations. However, the selective pressure for game evolution is found through optimizing a game to be interesting and fun for human players.

Building off recent progress in developing AI for card games [13], in this paper we seek to leverage computational techniques to understand evolutionary selection pressures on games. In particular, we compare and contrast the player experience through calculating heuristics [6] across various versions of the card game HEARTS. Previously used as a testbed for AI research by Sturtevant et al. [18], HEARTS is a simple popular trick avoidance game with a long history, where the player with the most points at the end of the game loses. For consistency, we employ a general game playing approach [9], applying the same AI algorithm across multiple games, to computationally understand the heuristic ramifications of each variant.

Specifically, we pose the following questions related HEARTS:

1. How do changes in the rules manifest in player experience?
2. What is the relationship between ludemic space and heuristic space?

Supported by Hendrix Odyssey Summer Research Grants.

C. Browne et al. (Eds.): ACG 2021, LNCS 13262, pp. 247–256, 2022.
https://doi.org/10.1007/978-3-031-11488-5_22

We begin with a brief explanation of Hearts and its suitability for analysis, followed by a discussion of ten diverse rule variants. We next develop four heuristic metrics to help understand the player experience of card games, and then apply these metrics to each of the variants through computational simulation in CardStock [3]. Finally, we analyze the computational player's experience for each variant, cluster the variants to better understand their differences heuristic space, and conclude with avenues for future work.

2   Hearts

Typically, a game of HEARTS is played over multiple rounds until one player accumulates 100 points. To standardize our analysis and increase the speed of our simulations, each variant of HEARTS analyzed will consist of only one round with exactly four players, with no passing of cards between players. The current canonical rules of Hearts [14] can be summarized as follows:

> A one-round game of HEARTS for four players consists of thirteen tricks. First, shuffle a standard deck of cards. Each player receives thirteen cards. For each trick, players play one card to the trick. The first player will set the lead suit for the trick, which subsequent players must follow suit if they can, otherwise they may play any card from their hand. Also, the first player is restricted to not play a card from the Hearts suit unless one has already been played. Once all cards have been played, the player who played the highest card that matches the suit of the led card will collect all the cards in the trick and become the first player for the next trick. Once all tricks have been played, players earn one point for each $\heartsuit$ collected in tricks, plus 13 points if they collected the Q$\spadesuit$. If a player happens to collect all $\heartsuit$ and the Q$\spadesuit$, then they will *Shoot the Moon* and instead subtract 26 points from their score. The player with the lowest point value wins the game.

Multiple ludemes make HEARTS distinctive from other trick-taking games. First, the goal is to avoid taking tricks that contain certain cards instead of accumulate them. Players must avoid the whole suit of $\heartsuit$, but the Q$\spadesuit$ is the most critical to avoid because of its high point value. In addition, the normal restriction where players must follow the led suit in a trick is compounded with a new limit that players must *not play* $\heartsuit$ until there is no other option. Finally, players have the ability to recover from initial poor play through by collecting every point and reverse their situation to a winning position.

3   Variants

David Parlett describes the history of HEARTS and a multitude of variants to the basic rules [14]. For our analysis, we examine ten specific modifications to the standard rules given in Sect. 2. Table 1 summarizes the specific rule changes for each variant we examined. Figure 1 organizes these variants in terms of what

Table 1. Variants of Hearts and their attributes

Variant	First	Last	♡ Broken	Moon	Deck	Points
SlobberHannes	✓	✓			32	Q♣:1
Polgnac	✓	✓			32	J♡:1, J◇:1, J♣:1, J♠:2
Pure Hearts					52	♡:1
Black Lady					52	♡:1, Q♠:13
Black Maria					52	♡:1, Q♠:13, A♠:10, K♠:7
Broken Hearts			✓	.	52	♡:1, Q♠:13
Hearts			✓	✓	52	♡:1, Q♠:13
Grey Lady			✓	✓	52	♡:1, Q♠:7
Omnibus Hearts			✓	✓	52	♡:1, Q♠:13 J◇:-10
Spot Hearts			✓	✓	52	♡:X, Q♠:13
Widow Hearts		✓	✓	✓	51	♡:1, Q♠:13

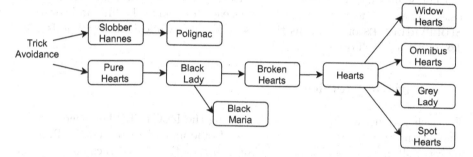

Fig. 1. Evolutionary History of Hearts Variants and Relatives.

is currently known about their historical progression through changes via edits, insertions, and deletions to the canonical HEARTS rules in ludemic space [5].

To examine historical ancestors of HEARTS, we start with bare-bones PURE HEARTS which has only 1 point for each ♡ collected and no rules for breaking ♡ or to *Shoot the Moon*. BLACK LADY adds in the 13 points for collecting the Q♠, and its offshoot BLACK MARIA adds additional 10 points for the A♠ and 7 for the K♠. We denote the breaking ♡ restriction as BROKEN HEARTS, and when the *Shoot the Moon* scoring is added, we arrive at modern HEARTS.

Many variants to modern HEARTS can be created by making small mutations which introduce alternate methods of scoring points via collected cards. For instance, GREY LADY reduces the points for Q♠ to only 7 points, while OMNIBUS HEARTS shifts in the other direction to make the J◇ worth -10 points. In SPOT HEARTS, each ♡ is worth its pip value rather than 1 point. Parlett states that these variants attempt to mitigate the large point value of the Q♠.

One additional variant is WIDOW HEARTS in which the 2♠ is removed, each player is only dealt 12 cards, and the 3 leftover cards are collected at the end of the game by the player who wins the last trick. We also examine two distantly

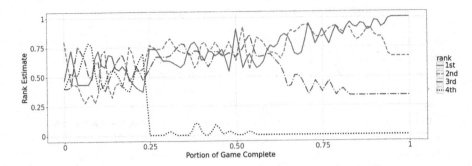

Fig. 2. Lead history for a typical game of Hearts.

related historical cousins in this vein, SLOBBERHANNES and POLIGNAC. Both of these games use a smaller 32 card deck, deleting all cards with rank 6 and below, as well as adding 1 point penalties each for taking the first and last trick. SLOBBERHANNES only assigns the Q♠ one point, while POLIGNAC adds 2 for J♠, and 1 for each other Jack.

4   Heuristic Metrics

For each variant, we encoded the rules using the RECYCLE language and ran simulations in CardStock with a mixture of random and AI players [3]. Random players will make choices using a uniform distribution across each choice, while the AI players will use statistics gathered from random simulations for each choice to determine their best chance of winning.

Our AI players employ a Perfect Information Monte Carlo (PIMC) strategy [11]. When faced with a choice, for each potential move, an AI player creates 10 clones of the current game state, mapping identically all known information from the player's perspective (cards previously played plus cards in their own hand), and creating a random determinization of the hidden information (cards in other player's hands) [19]. In the clone, all players are assigned to make choices randomly. The clone is played out to completion, and each player is assigned a value based on their final rank, scaled so that 1st place is mapped to 1, and 4th place is mapped to 0. In the event of tied ranks, all tied players earn the higher rank. These ranks are accumulated and averaged across all clones, and the move where the AI player earned the highest rank is selected.

To understand of the shape and flow of player experience, we examine *lead histories*, which record the rank estimates for all players every time any AI player makes a move. A lead history will have two dimensions, the estimated player *rank*, and the number of *moves* in the game. As an example, Fig. 2 shows the AI estimates of player rank in a typical 4 player game of Hearts. In the beginning of the game, most players are estimated to be in the middle with a good chance of winning. However, we can see a critical point about one quarter

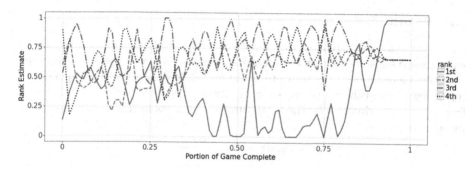

Fig. 3. Lead history demonstrating an AI player winning by *Shooting the Moon.*

of the way through the game where it is clear to everyone the losing player has lost and can never recover.

From these lead histories, we are able calculate four heuristic metrics, adapted from work by Browne and Maire [6]. Each heuristic is calibrated to a 0–1 scale, with 0 being no evidence of this quality, and 1 being high evidence.

4.1   Drama

If a player can come from behind and eventually win a game, we label this as dramatic. Figure 3 shows a dramatic run of Hearts where one player collects all the point cards to *Shoot the Moon.* We can see the winning player only solidifies their win right before the last trick of the game. We observed this behavior in 10% of our simulations.

We define *drama* as the average severity of being in a trailing position for the eventual winner. First, we define *dthresh*, a threshold for drama, set between the top ranked player and the next highest rank, so that when a player estimates their rank above the threshold, they believe it is more likely than not that they will be the winner of the game. Using np to denote the number of players,

$$dthresh = \frac{1 + \left(\frac{np-2}{np-1}\right)}{2} \tag{1}$$

In a two-player game, the threshold will be 0.5, half-way in-between the winning and losing ranks of 1 and 0 respectively. In a four-player game, the threshold will equal 5/6.

The full *drama* heuristic is then calculated using the winning player's path through the lead history. The number of times the leader is below the threshold is *dcount*. Each time their estimate falls below the drama threshold, the difference between the threshold and the estimate is calculated, and the sum of these differences is averaged. These differences are also scaled with the square root, so that larger differences are weighted more heavily in the final average.

$$drama = \frac{\sum_{i=1}^{dcount} \sqrt{dthresh - est_{winner,i}}}{dcount} \tag{2}$$

4.2 Security

One other calculation related to drama is the notion of the lead security of the winning player. A simple way to determine security is the percentage of the game that the winner was in the lead in the game. While drama can be impacted by just a few poor evaluations, the security heuristic is more stable. Using *dcount* from above, and dividing by the total number of *moves* in the game gives us the following equation.

$$security = 1 - \frac{dcount}{moves} \tag{3}$$

4.3 Spread

When deciding which move to make, an AI will try to determine their chances of winning for each given move. As a player looks at their possible moves in the game, many times they can identify some moves quickly as good and others as bad. Other times, it is difficult to know which move will have the best outcomes. If there is a difference in the win percentage estimates between possible moves, then this is a meaningful choice for the player: they should choose the move that gives them the best estimate. If there is no difference, then the move is meaningless.

By subtracting the minimum estimate from the maximum estimate (which will ultimately be chosen by the player) at each turn, we can calculate the spread at choice i (s_i) between these moves. If we find consistently high spread throughout the whole game, this will indicate that the game is a series of interesting decisions [1].

If we define the number of choices a player has in the game as $|choices|$ then we can determine the degree to which a player has meaningful moves by:

$$spread = \frac{\sum_{i=1}^{|choices|} s_i}{|choices|} \tag{4}$$

4.4 Order

Finally, we wish to determine how much control a player has over their own fate, or if they are at the whims of random events. When the *order* is low, this indicates the AI player has a hard time winning against random chaotic players, but when it is high, the AI player is very successful in determining their success in the game. To calculate the *order* heuristic, first, we record the win percentage ($aiwp$) of the AI player in games with one AI player and the rest Random players. The AI player is always goes first.

Next, we determine the expected win percentage (ewp) for the number of players in the game, assuming that the game is fair. For our games with 4 players, a fair game would expect the first player win 25% of the time. A perfect AI in a perfect information world should be able to win 100% of the games. This is reduced as unaccounted for chaos through hidden information is introduced.

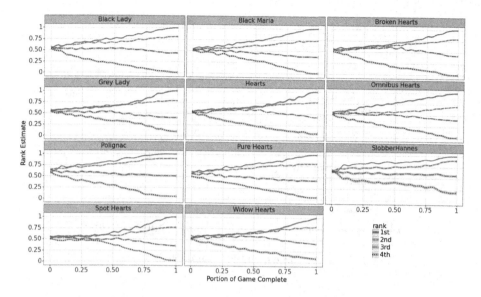

Fig. 4. Averaged lead histories for Hearts and each of the ten variants.

Therefore, we can calculate the *order* of the game by finding the ratio of the *aiwp* gain over the perfect ordered AI gain as follows:

$$order = \frac{aiwp - ewp}{1 - ewp} \tag{5}$$

5   Results

To gather statistics for each variant in CardStock, we ran 100 games with one AI versus three random players, plus 100 games with all AI players. Figure 4 shows the averaged lead histories for each variant. These were calculated by averaging for each rank across the 100 simulations with all AI players.

First, we can trace the effects of historical progression, starting with PURE HEARTS. The addition of the Q♠ in BLACK LADY has a significant impact on the fortunes of the losing player early in the game, steepening their decline. Looking next to BROKEN HEARTS, Parlett states that the hearts-breaking restrictions "feel unnecessary", [14], however in our simulations, this variant has the clear effect of delaying the separation of the top three players until the midgame. On these graphs, there appear to be no large differences when adding in the *Shoot The Moon* rule.

When comparing the point-focused modern variants, it is clear that SPOT HEARTS and GREY LADY are the most effective at mitigating the Q♠, with SPOT HEARTS pushing any separation between player's expected ranks until the midgame. OMNIBUS HEARTS with its reward for the J◊ gives the winning

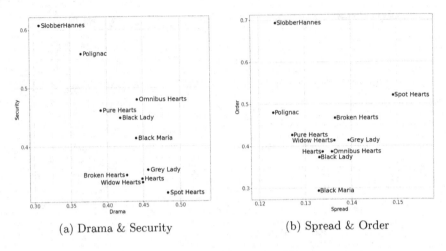

(a) Drama & Security   (b) Spread & Order

Fig. 5. Heuristic space quantification of Hearts variants.

player some early separation from the pack, while WIDOW HEARTS delays the final rank determination until the last trick.

We also see a drastic difference in SLOBBERHANNES and POLIGNAC. It appears the penalty for taking the first trick has a direct impact on the fate of the losing player. Also, due to SLOBBERHANNES having only one penalty card, there is not much variety for all players until the last trick.

Figure 5 shows each variant plotted on each of the metric dimensions of *drama*, *security*, *spread* and *order*, with the pairing chosen to facilitate ease of visualization of the results. First, we can see evidence of evolutionary pressure on HEARTS towards higher *drama* and lower *security*. SLOBBERHANNES and POLIGNAC, two early variants, score high on *security* and low on *drama*. The core variants of PURE HEARTS and BLACK LADY fall in the middle of this graph, while SPOT HEARTS, a modern variant, has highest *drama* and least *security*.

We also find evidence of selective pressure toward higher *spread* scores and more interesting variants, where those variants with more diverse point structure such as SPOT HEARTS, GREY LADY, and OMNIBUS HEARTS, tend to have larger *spread*. When comparing variants on *order*, there is a less direct connection. However, once again we see SLOBBERHANNES is an outlier, where players have the highest chance to win against random players.

SPOT HEARTS appears to have many appealing qualities, however, it is not the current dominant variant. We believe this is because for humans, HEARTS is meant to be a light game, and time spent calculating the score with each ♡ worth different points is too high when compared to the simple 1 point per ♡ math, and this illustrates a limitation of our computational approach.

Finally, Fig. 6 shows a clustering of these variants using the four heuristic metrics described above. We normalized each metric dimension to a range of 0–1, calculated the distance matrix between all variants, and derived a hierarchical clustering using UPGMA [17]. When compared to Fig.1, the uniqueness of

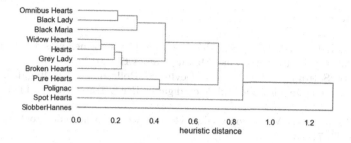

Fig. 6. UPGMA Clustering of Hearts variants in heuristic space.

SLOBBERHANNES and POLIGNAC is clearly present in both ludemic and heuristic space. SPOT HEARTS, which we noted previously was on the other end of the spectrum from these two cousins, is also found to be quite distant from the other variants, enough to provide a distinct play experience. The closest to HEARTS are WIDOW HEARTS, GREY LADY, BROKEN HEARTS, which also matches their underlying ludemic distance.

6 Future Work

There are a number of opportunities for improvement on our current work. While the PIMC players we used can make intelligent decisions, as shown by their ability to *Shoot the Moon,* they are very simple in comparison to more advanced AI methods such as Information Set Monte Carlo Tree Search (ISMCTS) [19] or Conterfactual Regret Minimization (CFR) [4] available through OpenSpiel [10]. Also, we limited our analysis of HEARTS variants to one round, however as Neller and Presser [12] demonstrate for the simple dice game Pig, optimal strategy changes when rounds are played within a full game, and we believe the same will hold for HEARTS. In addition, the number of heuristics we calculated is small and only gives a window into the full player experience. A richer set of heuristics will provide a larger space for more accurate clustering.

Looking forward, the analysis presented here is easily extendable to other player counts beyond four, and could be used to determine if a game retains the same heuristic qualities under different numbers of players. Finally, we envision creating a full map of the heuristic space of card games, including related trick-taking games such as SPADES [2], DOPPELKOPF [16], and SKAT [7], as well as different genres such as shedding, fishing, or press-your-luck. We believe that this map will illuminate connections across families and assist players in finding games suited to their individual taste.

Acknowledgements. The authors thank Anna Holmes and Daniel Sweeney for their contributions to the CardStock project, and the insightful comments of our anonymous reviewers.

References

1. Alexander, L.: GDC 2012: Sid Meier on how to see games as sets of interesting decisions. Gamasutra. Art. Bus. Making Games **f7** (2012)
2. Baier, H., Sattaur, A., Powley, E.J., Devlin, S., Rollason, J., Cowling, P.I.: Emulating human play in a leading mobile card game. IEEE Trans. Games **11**(4), 386–395 (2018)
3. Bell, C., Goadrich, M.: Automated playtesting with recycled cardstock. Game Puzzle Des. **2**, 71–83 (2016)
4. Brown, N., Lerer, A., Gross, S., Sandholm, T.: Deep counterfactual regret minimization. In: International Conference on Machine Learning. pp. 793–802. PMLR (2019)
5. Browne, C.: AI for ancient games. KI-Künstliche Intelligenz **34**(1), 89–93 (2020)
6. Browne, C., Maire, F.: Evolutionary game design. IEEE Trans. Comput. Intell. AI Games **2**(1), 1–16 (2010)
7. Buro, M., Long, J.R., Furtak, T., Sturtevant, N.R.: Improving state evaluation, inference, and search in trick-based card games. In: IJCAI, pp. 1407–1413 (2009)
8. Engelstein, G., Shalev, I.: Building Blocks of Tabletop Game Design: An Encyclopedia of Mechanisms. CRC Press, Boca Raton (2019)
9. Genesereth, M., Love, N., Pell, B.: General game playing: Overview of the AAAI competition. AI Mag. **26**(2), 62–62 (2005)
10. Lanctot, M., et al.: OpenSpiel: a framework for reinforcement learning in games. CoRR abs/1908.09453 (2019). http://arxiv.org/abs/1908.09453
11. Long, J., Sturtevant, N., Buro, M., Furtak, T.: Understanding the success of perfect information Monte Carlo sampling in game tree search. In: Proceedings of the AAAI Conference on Artificial Intelligence. vol. 24 (2010)
12. Neller, T.W., Presser, C.G.: Optimal play of the dice game Pig. The UMAP Journal **25**(1) (2004)
13. Niklaus, J., Alberti, M., Pondenkandath, V., Ingold, R., Liwicki, M.: Survey of artificial intelligence for card games and its application to the Swiss game Jass. In: 2019 6th Swiss Conference on Data Science (SDS), pp. 25–30. IEEE (2019)
14. Parlett, D.: A History of Card Games. Oxford University Press, USA (1991)
15. Parlett, D.: What's a ludeme? Game Puzzle Des. **2**(2), 81 (2017)
16. Sievers, S., Helmert, M.: A Doppelkopf player based on UCT. In: Hölldobler, S., Krötzsch, M., Peñaloza, R., Rudolph, S. (eds.) KI 2015. LNCS (LNAI), vol. 9324, pp. 151–165. Springer, Cham (2015). https://doi.org/10.1007/978-3-319-24489-1_12
17. Sokal, R.R.: A statistical method for evaluating systematic relationships. Univ. Kansas, Sci. Bull. **38**, 1409–1438 (1958)
18. Sturtevant, N.R., White, A.M.: Feature construction for reinforcement learning in hearts. In: van den Herik, H.J., Ciancarini, P., Donkers, H.H.L.M.J. (eds.) CG 2006. LNCS, vol. 4630, pp. 122–134. Springer, Heidelberg (2007). https://doi.org/10.1007/978-3-540-75538-8_11
19. Whitehouse, D., Powley, E.J., Cowling, P.I.: Determinization and information set monte carlo tree search for the card game dou di zhu. In: 2011 IEEE Conference on Computational Intelligence and Games (CIG'11), pp. 87–94. IEEE (2011)

Author Index

Printed in the United States
by Baker & Taylor Publisher Services